THEOPHRASTUS OF ERESUS

SOURCES FOR HIS LIFE, WRITINGS THOUGHT AND INFLUENCE

COMMENTARY VOLUME 2

LOGIC

PHILOSOPHIA ANTIQUA

A SERIES OF STUDIES ON ANCIENT PHILOSOPHY

VOLUME CIII

THEOPHRASTUS OF ERESUS

SOURCES FOR HIS LIFE, WRITINGS
THOUGHT AND INFLUENCE

COMMENTARY VOLUME 2
LOGIC

PAMELA HUBY

THEOPHRASTUS OF ERESUS

SOURCES FOR HIS LIFE, WRITINGS THOUGHT AND INFLUENCE

COMMENTARY VOLUME 2

LOGIC

BY

PAMELA HUBY

WITH CONTRIBUTIONS ON THE ARABIC MATERIAL

BY

DIMITRI GUTAS

TUTA SUB AEGIDE PALLAS · 1683 ·

BRILL

LEIDEN · BOSTON

2007

This book is printed on acid-free paper.

Library of Congress Cataloging-in-Publication Data

A C.I.P. record for this book is available from the Library of Congress.

ISSN 0079-1687
ISBN-13: 978 90 04 15298 4
ISBN-10: 90 04 15298 9

PRINTED IN THE NETHERLANDS

In Happy Memory of Bill and Martha Kneale

CONTENTS

PREFACE

This volume contains commentary on the sections concerned with logic (texts **68–136**) of the collection of texts published in 1992 (*Theophrastus of Eresus: Sources for his Life, Writings, Thought and Influence*, edited and translated by William W. Fortenbaugh, Pamela M. Huby, Robert W. Sharples (Greek and Latin) and Dimitri Gutas (Arabic) and five others, 2 vols., Leiden: Brill, 1992). It was comparatively easy to isolate those texts connected with logic, though in a few cases there was uncertainty about whether an item was to be assigned to rhetoric rather than logic. There was also little difficulty with problems of texts where the attribution to Theophrastus is doubtful.

The texts on which we are commenting are nearly all ones that contain the name of Theophrastus, along with a few in which only "the colleagues of Aristotle" are mentioned in a context where it is clear that Theophrastus is intended, usually with Eudemus. They are evidence for works now lost, even in translation. We have taken account, either by actual quotation or by giving references in the upper apparatus, of all such passages up to the cut-off date of 1450. Two items printed in the appendix are without attribution, and are included only as possibly by Theophrastus. To facilitate access to contexts we have added references to English translations of some passages quoted or referred to, and have given short accounts of most of the items mentioned in the upper apparatus.

Within the commentary in some cases several items are grouped together for a general discussion, but then individual items are also treated separately. Lists of relevant literature are given either under the heading of a group or with individual items. We have transliterated short items of Greek, but quoted longer ones in the original script.

Many people have helped me in this work. Jonathan Barnes and John Faris read parts of it and saved me from many mistakes; those that remain are my own. I am also indebted to the late Bill and Martha Kneale who discussed many knotty points with me; the late Hans Gottschalk let me see some of his unpublished material; I have benefited from discussion with or advice from Jacques Brunschwig, Myles Burnyeat, Bill Fortenbaugh, without whom this whole enterprise would never have come about, the late A.C. Lloyd, the late Mario Mignucci, David Sedley, Bob Sharples, and Fritz Zimmermann. In addition, Priscilla Huby, Adam Huby, and Timothy Havenhand helped me about technical matters. In this volume the work of Dimitri Gutas has been of especial value because in some areas the Arabic material fills important gaps in our evidence. Finally I must mention the late George Kerferd for his continual support.

It was only after the bulk of this work had been written that I became aware of the important study of De Rijk, entitled *Aristotle Semantics and Ontology*, which in fact contains a great deal of valuable work on Aristotle's logic. I have however been able to incorporate many references to it, either in the text or in footnotes.

ABBREVIATIONS

ACOA:	Ancient Commentators on Aristotle London: Duckworth.
ANRW:	*Aufstieg und Niedergang der römischen Welt.*
BICS:	*Bulletin of the Institute of Classical Studies.*
BT:	*Bibliotheca Teubneriana*, Leipzig: B.G. Teubner.
CAG:	*Commentaria in Aristotelem Graeca*, Berlin: Reimer, 1882–1909.
CMG:	*Corpus Medicorum Graecorum*, Leipzig: Academia Berolinensis, etc., 1908.
CQ:	*Classical Quarterly.*
DPA:	*Dictionnaire des Philosophes Antiques*, ed. R. Goulet and M.O. Goulet-Cazé, Paris 1989ff.
F:	Ibn-an-Nadīm, Kitāb al-Fihrist, Gustav Flügel, Johannes Roediger, August Müller (eds.) 2 vols., Leipzig 1871–2.
FHS&G:	See Preface p. ix.
LCM:	*Liverpool Classical Monthly.*
LSJ:	H.G Liddell and R. Scott, rev. H.S. Jones, *A Greek-English Lexicon*, ninth edition, Oxford: Clarendon Press, 1940.
OSAP:	*Oxford Studies in Ancient Philosophy*, Oxford: Clarendon Press.
PCSB:	*The Philosophy of the Commentators 200–600 AD.* ed., R. Sorabji, London: Duckworth 2004.
PL:	*Patrologia Latina* ed., J.P. Migne, Paris 1844–1904.
RE:	*Paulys Real-Encyclopädie der Altertumswissenschaft*, ed., K. Wissowa, Stuttgart: Metzler/Munich: Druckmüller 1894–1978 (with supplements).
Rh. Mus.:	*Rheinisches Museum für Philologie.*
RUSCH:	*Rutgers University Studies in Classical Humanities.*
AlexPA:	Alexander of Aphrodisias, *On Aristotle's* Prior Analytics.
AnonTarán:	Anonymous, *On Aristotle's* De interpretatione (cod. Paris. Gr. 2064 Tarán).
DI:	*De interpretatione.*
PA:	*Prior Analytics.*
PostA:	*Posterior Analytics.*

The following works are cited by author(s)' name. For full details see Bibliography.

Ackrill:	Aristotle's *Categories* and *de Interpretatione.*
Bocheński:	*La Logique de Théophraste.*
Flannery:	*Ways into the Logic of Alexander of Aphrodisias.*
Graeser:	*Die logischen Fragmente des Theophrasts.*
Kieffer:	Galen's *Institutio Logica*, English Translation, Introduction, and Commentary.
Kneales:	*The Development of Logic.*
Lukasiewicz:	*Aristotle's Syllogistic.*
Prantl:	*Geschichte der Logik im Abendlande.*
Regenbogen:	'Theophrastos', *RE* suppl.7 (1940).
Repici:	*La logica di Teofrasto.*
Ross:	Aristotle's Prior and Posterior Analytics.
Wehrli:	*Die Schule des Aristoteles.*
Zeyl:	*Encyclopedia of Classical Philosophy.*

Translations of many of the commentators on Aristotle are now available in the series edited by Richard Sorabji (*ACOA*). They include:

AlexPA (Barnes):	Alexander of Aphrodisias, *On Aristotle's* Prior Analytics 1.1–7.
AlexPA (Mueller)[1]:	Alexander of Aphrodisias, *On Aristotle's* Prior Analytics 1.8–13 etc. trans. Ian Mueller with Josiah Gould.
AlexPA (Mueller)[2]:	Analytics 1.14–22 trans. Ian Mueller with Josiah Gould.
AlexPA (Mueller)[3]:	Alexander of Aphrodisias, *On Aristotle's* Prior Analytics 1.23–31 trans. Ian Mueller.
AlexPA (Mueller)[4]:	Alexander of Aphrodisias, *On Aristotle's* Prior Analytics 1.32–46 trans. Ian Mueller.
AlexMet (Madigan):	Alexander of Aphrodisias, *On Aristotle's Metaphysics* 4 trans. Arthur Madigan.
AlexMet (Dooley):	Alexander of Aphrodisias, *On Aristotle's Metaphysics* 5 trans. W.E. Dooley.
AlexTop (VanOp):	Alexander of Aphrodisias, *On Aristotle's* Topics 1 trans. J.M. Van Ophuijsen.

AmmonBlank: Ammonius, *On Aristotle's* On Interpretation 1–8 trans. David Blank.
SimpCat (Chase): Simplicius, *On Aristotle's* Categories 1–4 trans. Michael Chase.

DL: = Diogenes Laertius, *Lives of the Philosophers.*

INTRODUCTION

Theophrastus introduced a number of innovations into Peripatetic logic, sometimes independently but, in the case of modal logic, working with his colleague Eudemus. Mostly he followed in the footsteps of Aristotle, and several of his works have the same titles as those of the latter. But in places he was critical of Aristotle, as in modal logic, and in others he added to what Aristotle had done, as on hypothetical syllogisms.

It will be seen that in some areas modern scholars have differed widely about the independence and the value of his work. His innovations in modal logic have generally been seen as improvements on the pioneering efforts of Aristotle, a view held even in ancient times. It is interesting, however, that for some aspects of that field we are dependent on Arabic evidence, because the works in Greek on which Averroes and others drew are lost to us.

More controversial has been the claim that he had a strong influence on the development of Stoic logic: some have held that there was no connection between Peripatetics and Stoics at the beginning, and that the two schools followed their own lines, although later thinkers blurred the distinctions and developed a common logical vocabulary. Others, however, find it difficult to believe that two groups living in Athens at the same time could fail to influence one another.

At the other end of our period the ideas of Theophrastus, having passed through a number of intermediaries, had considerable influence in the Middle Ages. Bocheński 127 points out that he affected "classic" logic in several ways—the five extra moods, modal logic, and aspects of hypothetical syllogism. Inevitably he played second fiddle to Aristotle, but he still ranks high in the world of logic.

LOGIC

68–136 Regenbogen 1380–9. Graeser. Repici.

Overview: Sources

Alexander of Aphrodisias (c. 200 AD) was the greatest commentator on Aristotle and wrote on many of his works. Details of his life are sketchy, but it is likely that he taught in Rome. The first book of his Commentary on the *PA* survives, and is a major source for us, and we also have his work on the *Topics*. The survival of these works and the loss of the rest of his commentaries on the Organon probably account for the fact that we have more about Theophrastus' work on various kinds of syllogism than we do about any other branch of logic, though we do have a fair amount about his *On Assertion and Denial*, connected with Aristotle's *DI*. Other lost works of Alexander provided information for Arabic writers, especially Averroes. See Flannery xxi–xxiv for a view about how Alexander regarded Theophrastus.[1]

From **Themistius** (c. 317–88) we have a long passage, translated into Arabic, from his attack of Maximus about the reduction of syllogistic figures. We also have significant fragments from his paraphrase on *Prior Analytics*, also in Arabic (and in Hebrew translation from the Arabic) (DG) **Diogenes Laertius** (third century AD) has given a long list of Theophrastus' works which include many on logic. It is clear that many of these were linked with works of Aristotle, some sharing the same title. **Galen** (129–c. 210–15 AD), like many medical men, took an interest in logic, and wrote about several Peripatetic works. But he tells us comparatively little of substance about Theophrastus. He does however in places refer to "the ancients" in contexts where the views may be those of Theophrastus. Here we may also refer to **Apuleius** (mid-second century AD), the reputed author of a *De Interpretatione*; even if it is not his, it is still our earliest Latin work on logic. See on **92**.

[1] Moraux (1984) vol. 2 discusses many of the lost commentators of whom we know something through Alexander. Many of the logic sources are covered more fully, as being also connected with rhetoric, in Fortenbaugh 2005.

A number of later commentators claim knowledge of Theophrastus. **Olympiodorus** (c. 500–after 565) and **Philoponus** (c. 490–570s) were both students of **Ammonius** (c. 440–517/26), and **Elias** was probably a student of Olympiodorus, though Wildberg (1990) 38 thinks the evidence for this inadequate. We must also take account of **pseudo-Ammonius**, probably of the sixth century, whose work appears in Wallies' edition after that of the genuine Ammonius, the scribe having ended on p. 38 with instructions to look out for the continuation. Wallies has added a series of scholia from ff. 226^r–261^v which start with chapter 8. He takes them to be by a follower of Ammonius, and refers to the remark (p. 40.16) about "the man who explained this to me, following the *hupomnema* of Iamblichus". For **David** see on **71B**. **Simplicius** (c. 490–560), a very reliable witness, gives us disappointingly little. **Boethius** (c. 480–524/6) stands alone: there is still uncertainty about where he lived and worked and the sources of his information.

Some material survives only in Arabic, especially that of **Avicenna** (d. 1037) and **Averroes** (1126–97) on modal logic. It has also seemed right to refer to certain passages that could concern Theophrastus but do not actually name him or otherwise identify him.

There are some lost works which may have been intermediate sources. Thus, while **Porphyry**'s (c. 232–305) *Isagoge* and his *Commentary on the* Categories *in questions and answers* survive, his *Exegesis of the* Categories *to Gedalius* is lost, as is the set of commentaries which are referred to in **80**.

For more information about the commentators see Blumenthal 7–71. Although he concentrates on the psychological commentaries what he says is also of wider interest. *PCSB* also has much information about the commentators, both in the Introduction and under individual topics.

Repici 9–30 surveys editions and other material about the logical works until 1973.

Overview: Doctrines

Theophrastus worked within the field of Aristotelian logic, but developed some aspects of Aristotle's doctrines and challenged others. See **72**. Working with Eudemus he attacked Aristotle's unsatisfactory account of modal logic. Perhaps with Eudemus he made minor additions to the account of categorical syllogisms. He also developed the theories of prosleptic and hypothetical syllogisms, and may have recast the topics. It is likely that he was interested in reappraising Aristotle's material, e.g. about the order of categorical syllogisms (See **96AB**) and of the *Topics.*

The circumstances in which pieces of evidence for his views on logic have survived mean that there are many passages where we do not know whether what is quoted is an isolated comment, perhaps on something Aristotle had said, such as might have been made in a commentary, or part of a larger systematic development. Most of the logical works which are more than mere names to us are associated with Aristotle. This may be due to the fact that most of our evidence comes from commentaries which are themselves on Aristotle.

Repici 11–30 surveys studies of aspects of Theophrastus logic.

The language and symbolism to be used of Peripatetic logic

There is a variety of ways of dealing with the language used of syllogistic arguments by the ancient writers themselves, and a variety of later symbolisms to express them. In addition the individual moods were given nonsense names in the Middle Ages which are still used, and are useful, today. Since Aristotle himself was by no means consistent, and later Greek writers also differ, we decided to translate the Greek and Latin as closely as possible to the actual words used. And for the purpose of explaining Theophrastus it seemed unnecessary to go into the developments of his ideas worked out in interesting ways by modern logicians, though we hope to have given sufficient references to those developments for those who wish to follow them up.[2]

Friends have urged that it would help those who are not experts to have some information about the logical notions involved. We give an account of Aristotle's syllogistic at **91A–D**, but more should be added here.

First, there was a development over time in which Aristotle's usual way of setting out syllogisms was replaced by another in which the order of the terms was changed. Thus in the first syllogism Aristotle would have:

[2] For a very full account of Alexander's logical terms, which overlap extensively with those of Aristotle see AlexPA (Barnes) 18–32.

B belongs to all A	or	B is (said) of all A
C belongs to all B		C is (said) of all B
Therefore C belongs to all A		C is (said) of all A

Later this would become:

All As are B
All Bs are C
Therefore All As are C.

We use the following symbolisation:

AaB = All As are B
AeB = No As are B
AiB = Some As are B
AoB = Some As are not B

The middle term, which links the two premises, raises no difficulties, but Aristotle's treatment of what he calls the major and minor terms runs into trouble in the second and third figures. Various accounts of the "correct" position were given by later logicians, but it is best to accept that it cannot be done. See Lukasiewicz 28–30. Further, it seems unnecessary to follow Aristotle in using different letters for the terms in the second and third figures. It is also wrong to suppose that there is a correct order for the premises: the conclusion is the same whichever the order. See Lukasiewicz 32–4. Questions have also been raised about what Aristotle meant by a "perfect" syllogism. It seems to be no more than one which is self-evident and needing no proof. Questions can of course be asked about what is implied by this notion, but it is best to ignore them here.[3]

For a clear account of the medieval names of syllogistic moods see Kneales 232–3 and De Rijk (1967) 401–3. There is also a list with explanations in the Loeb translation of the *PA* Cambridge Mass.: Harvard University Press 1973 197.

Regenbogen 1380–3. Graeser 50–8. Repici 33–7. M. Sollenberger 'The Lives of the Peripatetics: An Analysis of the Contents and Structure of

[3] The Kneales 68–71 discuss the problems in Aristotle's account of major and minor terms.

Diogenes Laertius' "Vita Philosophorum" Book 5' in *ANRW* II 36.6 3849–55. Fortenbaugh (1998).

We should distinguish two questions, what, if any, was the title that Theophrastus gave to a work, and what titles appear in various references to it.[4] The former is difficult. Not even Aristotle seems always to have given a definitive name to his works: in particular we are lumbered with *De interpretatione* and its Greek equivalent which are improbable.[5]

Aristotle's papers were probably "edited" by Theophrastus and others, and we may assume that at some stage titles were assigned, but it is only when we come to what might be called library work that serious titling seems to have occurred. So we have "library lists" and we also have references to works they are writing about by e.g. Galen and Alexander. By their time, four centuries after Theophrastus, titles were standardised. But they were not librarians, though Galen did his best to list his own works.

As with Aristotle's works, there appear to be blocks of different origins in DL's list of those of Theophrastus. First (5.42–46) there is a long one in alphabetical order—in the sense that the order is that of the first letter of the first significant word. There are apparent exceptions, but they can all be accounted for. This is followed by a shorter section (5 46–8) also in alphabetical order. Then (5.48–50) come roughly thirty titles in a non-alphabetical list, followed by (5.50) roughly twenty again in alphabetical order. How these lists are related among themselves is hard to tell. There is some apparent overlap between them, but not a lot. Fortenbaugh has suggested that there are five separate lists bought separately by the library of Alexandria.

We are uncertain about the opening of the first list, where six logical titles appear. One might assume that they were put together because they were all on logic, but it happens that they all also begin with alpha. Other apparently logical titles are scattered through the lists.[6] 6a, 7a, 8, and 9a are all closely connected with Aristotle's *Analytics*, and 18a perhaps with that and his *Topics*. 26 stands alone.[7]

[4] See Stephen White (2002) 9–37 for a thorough study of DL's catalogue and a discussion of problems about titles. Logical titles are dealt with at 13–14.

[5] For a discussion of this title see Whitaker (1996) 5–7 and De Rijk (2002) I 190–1. Unusually, De Rijk thinks the title appropriate.

[6] Barnes (1985) 128 = Wiesner 561, says that DL's list is certainly defective, and there is no reason to suppose that we have a complete list of Theophrastus' logical works.

[7] The same first five are given by the Suda at the head of its short list of works by Theophrastus (our **2**), and that in turn was followed by Eudocia Augusta, *Violarium* no. 453 (p. 370 Flack). This was in fact Constantinus Palaeocappa. See T. Dorandi, 'Diogenes Laertius Vitae Philosophorum' *Phronesis* 45 (2000) 332.

1a *Categories*
b *Commentary on the Categories* (Arabic)

2 *Commentary on his own Categories*

These are all controversial and will be discussed at **71**.
Categories is not on Diogenes' list.

3a *On Affirmation and Denial*

See Sollenberger (1985) 49–50.

3b *On Affirmation and Denial* (Latin)
c *On Affirmation*
d *The Book of Affirmative and Negative (Propositions)* (Arabic title)
e *De Interpretatione* (Greek title)

4 *On Denial*

See Sollenberger (1985) 49–50.
3–4 are also controversial and will be discussed at **72**.

5 *Primary Propositions*

This is known only from DL's list. It stands out from other logical works by its great length: only no. 22, *Dialectical Arguments* or *Epicheiremes* is similar and both are said to have 18 books. One might guess that both were collections of material for students to use in arguing. Graeser 54 (on his VIII) suggests that 5 might be connected with the starting-points (ἀρχαί) of demonstration of e.g. *PostA* 1.2 71b20–23. Brunschwig, on the other hand, (1967) ciii n. 3, thinks it might be a catalogue of premises suggested by Aristotle in *Topics* 1.14 105a34–b18.

6a *Prior Analytics*
b *Prior Analytics* (Arabic title)

Graeser 52 notes that no passage from books 2 and 3 remains, probably because the second book of Alexander's *Commentary* is missing. Flannery

29 n. 73 doubts if Theophrastus wrote a passage-by-passage commentary on Aristotle. He suggests rather an easier account in simpler terms.

Ross notes that Aristotle himself refers to the *Analytics*, but oes not use "Prior" or "Posterior", terms first known to us from Alexander. By then they seem to be established titles, but Galen, Alexander's contemporary, in *On his own books* 118.5–8 says only that "*Prior*" and "*Second*" *Analytics* are almost universal in his time. See Huby 2002 89–90.

These titles clearly refer to Theophrastus' study of Aristotle's *PA*.

6c *Analytics*

This may cover either or both works.

7a *Posterior Analytics*
b *Second Analytics*

Similarly these refer to a work connected with Aristotle's *PostA*.

8 *Epitome of the Analytics*

We know only that this was listed by DL with other logical works. The *Suda* presumably got it indirectly. See Graeser 53 (on his IV) for references to sceptics who thought it an Alexandrian work.

9a *On the Analysis of Syllogisms*
b *The Analysis of Syllogisms*

These must be identical, as 9b comes from a list given by the *Suda* based on that of Diogenes.

9c *On the Solution of Syllogisms*

See on 21. 9c is in the same alphabetical section of DL's list as 9a, which suggests that it is a different work.

10 *On the Deciding of Syllogisms*

This may refer to the same work as 9ab, but is listed separately by DL. According to Alexander (our **97**) *On the Analysis of Syllogisms* contained the method of analysing syllogisms as opposed to 18b, *Arguments Reduced to Figures.*

11 *Definitions connected with the Language of Syllogisms*

This is clearly a logical work. Bocheński 28 suggests that it is connected with *PA* 1.1. See Graeser 52.

12 *Definitions*

13 *Regarding Definitions*

14 *Regarding Definitions* (or *Terms*)

These may or may not be logical works. We should look at the words involved: in 11 the word is ὁριστικά—things connected with ὅροι—not found elsewhere in our material; διορισμός is found especially in **102** in connection with an account of the problematic, and presumably means "definition" there; ὁρισμός is not found elsewhere in Theophrastus; ὅρος is used frequently as a) "term" and b) "definition", and once (**106G**.2) as "premise".[8]

15 *Divisions*

16 *On Differentiae*

These may be logical works, but could be scientific ones. Stephen White (2002) 13 n. 15 thinks that this title might be an alternative title for *Peri heterophonias ("On Differences of Voice").* See Sharples' Commentary (1995) 43 n. 137 on **350** nos. 3 and 4. It has been suggested by Moraux (1984) vol. 2 249–93 that Aspasius' statement (**533**) that Theophrastus and Eudemus had the same view about friendship based on superiority related to a division that was logical rather than ethical.

17 *Topics*

The work bearing this name is related to Aristotle's *Topics*. Averroes appears to quote from it at length, through the mediation of Themis-

---

[8] Even here Bobzien (2000) thinks it could mean "term".

tius, in his own *Middle Commentary on Aristotle's* Topics, from which we give a selection in **126** and **132**. That suggests that Theophrastus followed Aristotle closely. It has been suggested that Theophrastus' *Topics* occupied only two books because he reduced individual topics to hypothetical syllogisms, but see on **122–36**.

8a *The Reduction of Topics*
b *Arguments Reduced to Figures*

These may, but need not, be alternative titles for the same work. See on **97**. Repici 40, following Solmsen (1968) 65 n. 1, suggests that the papyrus fragment from Florence (our App. 2) may be related to this. Solmsen there says that Theophrastus "decided to recast the topics into the form of hypothetical syllogisms." That can only be a conjecture. See also Slomkowski 112 n. 62.

19 *Preliminaries to the* Topics

Graeser. Frede (1987). Gottschalk (1987). Baltussen (2000).

This is otherwise unknown. See Graeser 54–5, Frede 18, Gottschalk 249–52 and Baltussen 31 n. 3, who favours Bocheński's (29) view that this was an introduction to dialectic. DL (5.24) also lists a *Preliminaries to the Topics* by Aristotle, and it is of Aristotle's works that Simplicius, *On Aristotle's* Categories 379.8–12 says that Aristotle's *Categories* was seen by some, including Andronicus, as preliminaries to the *Topics*, and the same view was held by Ammonius, *On Aristotle's* Categories 14.18–20,[9] but Alexander *On Aristotle's* Topics 5.27–8 says that the first book of the *Topics* was also known as Πρὸ τῶν τόπων ("*Before the Topics*"). Porphyry *On Aristotle's* Categories 56.18–31 says that some writers called the *Categories Introduction to the Topics*, but argues against this and other titles. See Strange's translation of this work 31 n. 12.[10] Was then Theophrastus' work of the same name, if it was a separate entity, similar to one of Aristotle's, or independent?[11]

[9] Moraux gives other references as well.

[10] Moraux (1974) 273–4 thinks that the view that Aristotle's *Topics* should follow his *Categories* was known to Adrastus (second cent. AD), but that does not help us about Theophrastus. Nor does Whitaker's argument (see note 6) that the *DI* was a preliminary to the *Topics*.

[11] Repici has *pros* for *pro*. That is clearly a mistake.

20a *On How Many (Senses)*
b *On Many (Senses)*

See **73, 133** and **134**. These titles are not on DL's list.

Of the remaining titles to 37 some are clearly logical, but others are uncertain and have been put here only on the balance of probabilities.

21 *Solutions*

This may be the same as 9c, but Graeser 53, following Regenbogen 1382 points out that Aristotle has an *Eristic Solutions* in 4 books (DL 5.22) with which it might be connected.

22 *Epicheiremes*

This is followed, in DL's list, by 24, *Objections*, and may be connected with it. 22 is in eighteen books which looks like a collection of examples for student use, and may be for rhetorical purposes.

23 *Epicheiremes* in 2 books

This is from a part of DL's list later than 22.

24 *Objections*

This follows 22 in DL's list.

26 *Polemics* or *The Theory of Eristic Arguments*

This is listed by DL with other logical works. Its title is uncertain, and various suggestions have been made, but they agree in separating the first word, as a title, from the rest by means of "or". See Sollenberger (1985) 49–50.

30 *On Greater or Lesser*

This may be on major and minor terms as in *PA* 1.4–6. Cf.26b37. Graeser 53 thinks it might be part of a larger whole.

32 *On Simple Problems*

The word *diaporêma* occurs once in Aristotle at *PostA* 2.8 93b20, where Barnes translates it "puzzles". There Aristotle could be referring to a known group of problems, but our information is too limited for certainty.

33 *On the Liar (Paradox)* in three books

This can hardly be other than a study of the liar paradox known from other sources. The paradox was invented by Eubulides, and not known to Plato. But Aristotle knew it. Barnes (1999) 39 notes that in DL's list of Chrysippus' works (7.197) there is a *Solution according to the ancients addressed to Dioscorides* which, in the position in which it is found, must be connected with the Liar argument. It is reasonable to suppose that Chrysippus had read Theophrastus' work on this and written about it.[12]

34 *Theses* in 24 books

Fortenbaugh (2005) 83–7

These may have been for student use, but such theses might not have been (only) logical. They are likely also to have been rhetorical, at least. See **74–6, 89**, and **135**.

–7 We know nothing of these. 36, *Thesis on the Soul*, may have been psycho-logical. There is a similar uncertainty about the work called *Theses on the Soul*, one book, in DL's list of Aristotle's writings (5 24), but there the item is in a list of theses of various kinds, whereas Theophrastus' work is in an alphabetical listing.[13]

1 line 237 *Memoranda* of Aristotle or Theophrastus in DL's list is in an alphabetical section, so that we cannot tell its content from its context. Simplicius, *On Aristotle's* Categories 65.4–6 refers to *hupomnêmata*

[12] Bocheński (1951) 101–2 gives a survey of the various accounts and solutions of the problem.

[13] Fortenbaugh (1998) 198–200, (2003) 210–12, and now (2005) 83–7 has some interesting speculations about this work.

of Aristotle related to the *Categories*, and to another, called Τὰ περὶ τὴν λέξιν ("*Things connected with expression*"), thought spurious by some, but certainly of Aristotle's school. See after **71**, and Galen at **73**.10. H.B. Gottschalk, 'Currents of philosophical thought in the first two centuries AD' *Dodona*: Ioannina 1987 89 is helpful about *hupomnêma*. See also n. 23 and David Blank in AmmonBlank 2–3.

We have placed *On False and True* at **246** 5 under Metaphysics on the balance of probabilities, but it could have been put here. Regenbogen 1380 treats *On Experience* (**265** no. 10) and (1383) *On Causes* (**137** no. 9) as logical; he suggests that **503** (Stobaeus 1 89.2–5) on types of causes, might belong to the latter. Repici 34 includes as logical works *On Signs* (**137** no. 17), but that is now established as being on weather signs (See P. Cronin 'The Authorship and Sources of the Peri semeion attributed to Theophrastus' in Fortenbaugh and Gutas (1992) 307–45). We have placed *On False and True* at **246** no. 5 under Metaphysics and *What are the ways of Knowing?* at **265** no. 7 under Psychology. This has been seen as related to Aristotle's *Metaphysics* by Wimmer, to the opening of the *PostA* by Bocheński 28, and to *PostA* 2.19 by Prantl I 350. We do not know.

Schenkeveld (1998) 67–80 suggests that FHS&G **683** should be removed from rhetoric to logic. It was indeed a difficult editorial decision to make. See after **71**.

The *Megarikos* or *Megarian Dialogue* (**1** 129) is also in an alphabetical section. It is referred to by DL at 6.22 (**511**) in connection with Diogenes of Sinope, the Cynic, where the point appears to be an ethical one. DL (2.112) says that Diogenes of Sinope wrote a dialogue to (*pros*) Ichthyas the Megarian,[14] and at 6.41 that he was in Megara. Further, Stilpo of Megara (c. 360–280), an almost exact contemporary of Theophrastus, is said (DL 2.113) to have attracted pupils from both Theophrastus and Aristotle. He also wrote a dialogue called *Aristotle* (DL 2.120). There is some uncertainty about the views of the Megarians, but it is possible that Theophrastus touched on the logical aspect of their work.

666 no 16 *On Statement (of the case) and Narration* was considered possibly logical by Prantl, but we have assigned it to Rhetoric.

[14] David Sedley, 'Diodorus and Hellenistic Philosophy', *Proc. Camb. Philol. Soc.* n.s. 23 (1977) argues that a Megarian was not necessarily from Megara, but belonged to the "school" whose founder was from Megara. See also M.J. White, DIALECTICAL SCHOOL in Żeyl 177–8.

Logical Writings

69 Ibn-an-Nadīm, *The Index* 7.1 The Names of Natural Philosophers (p. 255.13–14 Flügel)

Müller (1872) 41. Gutas (1985) 82.

The note on the flyleaf of the volume referred to by Ibn-an-Nadīm (written in 989 AD) must belong to one of the numerous scholars active in Baghdad during the entire period of the Graeco-Arabic translation movement (ca. 770–1000 AD), when bibliographical knowledge of classical authors was quite detailed. Ibn-an-Nadīm himself refers to two Aristotelian commentaries by Theophrastus (see **71I**, **72B**), and it would appear that these are also the commentaries referred to in the note. Müller, in his translation of this passage from the *Fihrist* in Flügel's edition, inexplicably omits the name of Theophrastus (DG).

70 Theodorus the Metochite *Philosophical* and *Historical Miscellanies* (p. 90.12–91.1 Mueller and Kiessling)

Theodore the Metochite (1270–1332) was a Byzantine statesman and polymath whose works included long commentaries on Aristotle. His *Miscellanies* were 120 essays on many subjects. This passage tells us only that Theophrastus sometimes disagreed with Aristotle. See **49** for Theodore's views about philosophers. For a survey see Wilson (1983) 256–64. Some of his views are discussed in Ierodiakonou (2002).

It is unclear whether the *hautai* ("these") of line 6 refers only to the works of the Stoics or also to those of Theophrastus, but the latter did sometimes disagree with his master, and the tenor of the passage seems to call for the inclusion of his name as opposed to Aristotle.

*Extra text (***70.1***)*

Rhazes, *Doubts (aporiai) about Galen*, Introduction (Muḥammad Ibn-Zakariyā ar-Rāzī, *Aš-Šukūk ʿalā* Gālīnūs, ed. Mehdi Mohaghegh, Tehran 1372 HŠ/1993, 2.6–14)

Text:

As for him who reproaches me and calls me ignorant for raising these doubts [about Galen] and discussing them, I pay no heed to him nor do I consider him a philosopher, since he has thrown the practice of philosophers behind his back and adhered to the practice of the rabble, which consists of imitating the leaders and never objecting to them.

For the practice of those engaged in philosophy has never ceased to be holding the leaders in high esteem while being relentless in questioning them hard and never indulging them. Here is Aristotle, who says, "Truth and Plato are at variance, and though they are both dear to us, truth is dearer to us than Plato," and raises objections to him and contradicts him in most of his opinions. And here is Theophrastus contradicting Aristotle [even] in the most self-evident part of philosophy after geometry, logic. And here is Themistius also bringing to light his [Aristotle's] mistakes in numerous passages, to the point that on occasion he stops and wonders, saying, "I have no idea how the philosopher could have overlooked this issue which is obvious to the extreme."

Comments:

This excellent variation on the theme of *amicus Plato* comes from the pen of the great physician and philosopher Rhazes (d. 925 or 935), the single most independently minded thinker in medieval Islamic civilization,[15] in the introduction to his systematic *aporiai* book about Galen. Rhazes' remark about Theophrastus's opposition to some aspects of Aristotelian logic in all likelihood refers to the issues of modal logic, discussed below at **98–109**. (DG)

This passage was drawn to our attention by Fritz Zimmermann, whom we thank. (PMH)

71A–I M. Frede, 'Titel, Einheit und Echtheit der aristotelischen Kategorienschrift' in P. Moraux and J. Wiesner (edd.) *Zweifelhaftes im Corpus Aristotelicum* Berlin: De Gruyter 1983 22–5, translated in M. Frede (1987) 11–28. H.B. Gottschalk, 'Did Theophrastus write a Categories?' *Philologus* 131 (1987) 245–53.

[15] Rhazes came as close to an atheist position as anyone could in his days. See now the detailed study by S. Stroumsa, *Freethinkers of Medieval Islam*, Leiden: Brill 1999, 87–120, with references to further bibliography.

The passages in **71** are concerned primarily with the title and authenticity of the *Categories*, and come from the openings of commentaries on that work or on Porphyry's *Isagoge*. We have practically nothing that can be seen as an excerpt from a work of that name by Theophrastus, and it is unlikely that any of these writers actually knew such a work. There is indeed no direct evidence for the existence of Theophrastus' *Categories* before Ammonius (c. 500), in contrast with the position with regard to his other works related to Aristotle's logical writings.[16] At **127A** we note that Simplicius uses Theophrastus' *Topics* where he might have been expected to use a *Categories* if it existed.

But there is some slight corroboration for the existence of other *Categories* attributed to Aristotle in Simplicius *On Aristotle's* Categories 18.16–21 trans. SimpCat (Chase) 33–4. Simplicius says that Adrastus in his *On the order of the writings of Aristotle* says that there is traditionally included (*pheretai*) as by Aristotle another book (*biblion*), and that it too is short and concise in expression (*lexin*) and differs with small divisions[17] and begins: τῶν ὄντων τὰ μέν ἐστιν ὁμώνυμα etc. Simplicius is puzzled by the fact that it has the same number of lines, so how was it short etc.[18] That indicates that Simplicius had not seen it himself. Adrastus lived in the second century AD, and had a good reputation. Since Simplicius is also reliable, this part of the apparently extravagant story about multiple works claiming to be by Aristotle has some corroboration.[19] But that does not help us much about Theophrastus. I. Hadot (1990) vol. 1 138–60 studies at length the five extant commentaries on the *Categories*.

D The text of Porphyry is given at the beginning of **71C**, and commentators have asked both why he said "also" (*kai*) for the explanation of the *Categories* of Aristotle, and why he added "of Aristotle". This leads many to consider the claim that there were other works called "*Categories*".

---

[16] No fragments of a *Categories* by any of Theophrastus' colleagues exist either.

[17] Presumably chapter divisions. Chase translates: "it differs only by a few divisions".

[18] Moraux (1984) vol. 2 316 thinks the second *Categories* was a paraphrase written in the first century BC to make Aristotle easier for his readers.

[19] Many of the commentators have similar accounts of a story presumably deriving from Adrastus. It amounts to a claim that there were in the library at Alexandria other versions of Aristotle's *Categories* and *Analytics*. No mention is made of any other work. Scholars are said to have examined them and decided that one *Categories* out of two, and four *Analytics* out of what are usually said to be forty, were genuine. The opening words of the second *Categories* are also frequently given. I. Hadot (1990) vol. 1 144–60 gives French translations of many of the relevant passages.

71A Ammonius, *On Porphyry's* Isagoge p. 1.3 (*CAG* vol. 4.3 p. 26.11–17 Busse)

For Ammonius' life see AmmonBlank 1–6, and Blumenthal (1996) 39–42.

This passage is found in the MSS as part of Ammonius' commentary on Porphyry's *Isagoge* or *Introduction* to Aristotle's work. For this see Wildberg (1990) 34–7. Ammonius wonders why Porphyry adds the words "of Aristotle" to "the *Categories*". This leads him to say that other *Categories* existed, including one by Theophrastus. Busse athetized this section (26.9–17), probably rightly. In the preface he has noted how much repetition there is, and at 31.2–9 Ammonius explains the use of *kai* ("also") as being because a knowledge of genus and species is useful for the teaching of Aristotle's *Categories* but also for what he says later. But some have referred to Archytas' *Categories.*[20] But Ammonius says this is silly because Porphyry has said what else the knowledge is useful for.

71B David, *On Porphyry's* Isagoge p. 1.3 (*CAG* vol. 18.2 p. 102.1–9 Busse)

For David (6th cent. AD) see Westerink at **71C** and J.P. Mahé 'David l'invincible dans la tradition arménienne' Appendix II in Hadot (1990) 189–207. There is more about the matter of spurious works at 81–2 and at David *On Aristotle's* Categories (*CAG* vol. 18.2 128), which Mahé quotes at 200–201. David seems to have been a pupil of Olympiodorus who became confused with an Armenian of the same name. Westerink p. xvi supports the view that he was from Thessaly. See also Wildberg (1990) 39–42,44–5 and on the app. to **71F**. Gottschalk thinks that **71B**, **71C**, and pseudo-Ammonius *On Porphyry's* Isagoge p. 1.3 (cod. Paris Gr. 1973) discussed below, are derived from a faulty copy of **71E**.

pseudo-Ammonius *On Porphyry's* Isagoge p. 1.3 (Paris Gr. 1973 f.21v) in Rose, *Aristoteles Pseudepigraphus* p. 129,14–18).

This at first sight seems different: it has οὐδ' ὅλως συμβάλλεται to the *Categories* of Theophrastus and Eudems which should mean "contributes not at all". But it seems likely that it is a version of David's account or a misunderstanding of it. Its MS. contains Ammonius' commentary on the *Isagoge* used as a peg on which to hang other expositions, including that of David. For more see Busse in *CAG* 4.3 xviii–xx.

In the margin of the MS is an attribution to "Ammonius the Little".

---

[20] This refers to the still extant Περὶ τῶν καθόλου λόγων ("*On Universal Terms*") attributed in ancient times to Archytas, Plato's contemporary, but now known to be part of the pseudo-Pythagorean literature of the first century BC. See Th. Szlezák (ed.) *Pseudo Archytas über die Kategorien* Berlin/New York: De Gruyter 1972 14, 153–7.

71C pseudo-Elias, *On Porphyry's* Isagoge p. 1.3 (p. 63.13–14 and 20–5 Westerink)

Westerink (1961) 126–33. H.J. Blumenthal, 'Pseudo-Elias and the *Isagoge* Commentators' *Rh. Mus.* 124 (1981) 188–92

This work also passed under the name of pseudo-David; see Westerink 129–31 for the complicated relationships in the tradition between David and Elias. He thinks the confusion started in the school at Alexandria "where both taught and where copies of their lectures would circulate without any name". He also suggests that this is an early version by David himself of his commentary. Blumenthal could accept the view that pseudo-David/Elias preceded David. The author follows the standard lines of a commentary on Porphyry. Wanda Wolska-Conus, 'Stephanus d'Athènes et Stephanus d'Alexandrie. Essai d'identification et de biographie *Revue des Etudes Byzantines* 47 (1989) suggests that the true author of this work is Stephanus of Alexandria, but this view has not been generally accepted.

Anonymous on Porphyry's *Isagoge* p. 1.3 (cod. Laur. 4.2 f.17) in Rose *Aristoteles Pseudepigraphus* p. 129,20–1

The title seems to be "Philoponus, *On Porphyry's* Isagoge", but the text differs from our text of Philoponus.

Westerink (p. xiii) thinks that this man lectured at Constantinople and was perhaps a medical man. He puts him after Olympiodorus who died in or soon after 565, and before 716, perhaps 680.

This passage adds the name of "Cleinias" (presumably Phainias) to those who wrote *Categories.*

71D Abū-l-Farağ Ibn-aṭ-Ṭayyib, *On Porphyry's* Eisagoge p. 1.3 (ed. K. Gyekye, *Ibn al-Ṭayyib's Commentary on Porphyry's Eisagoge*, Beyrouth 1975, 48.21–24). The Arabic text was translated by K. Gyekye, *Arabic Logic. Ibn al-Ṭayyib's Commentary on Porphyry's Eisagoge*, Albany, NY 1979, 33

Abū-l-Farağ Ibn-aṭ-Ṭayyib (d. 1048), a Nestorian philosopher, physician, and theologian, was one of the last representatives of the Baghdad Aristotelians, the historically significant school whose most famous member was al-Fārābī. He wrote voluminously on philosophy and medicine, many of his works consisting of commentaries on or paraphrases and abridgments of the Greek originals. His commentaries on the *Eisagoge* and the *Categories*, both extant, in essence repeat the material in the

Greek commentaries of late antiquity. For this reason it is interesting to note that in this passage, his text is so worded as not to ascribe directly to Theophrastus a work by the title of *Categories*. He says, instead, that Porphyry's five predicables are useful "for *the books* of Plato, Pythagoras, Theophrastus," etc. (emphasis added), leaving the precise reference vague. Now Ibn-at-Tayyib is the only author among those listed under our entry **71** that does this. Since he is also known to have followed very closely his Greek sources—in essence we can be sure that he had access to these very commentaries listed in **71**—his deviation from them in this instance must indicate that he had doubts about the accuracy of the report. One can only assume that he could not find any reference to or a copy of such a work, as listed by Ibn-an-Nadīm (**3D**)—especially since Ibn-an-Nadīm also mentions that the ascription was doubtful—and decided to express himself in a more circumspect way. (DG)

71E–H These are concerned with the authenticity of Aristotle's *Categories*, but tell the same tale as above about the existence of other works with a similar title. In **71F** Olympiodorus (495/505–565 or soon after) is the most extravagant, and has Theophrastus even writing a commentary on his own *Categories*.

71E Philoponus, *On Aristotle's* Categories, Introduction (*CAG* vol. 13.1 p. 7.16–17 and 19–22 Busse)

For Philoponus' career and the meaning of his name see Blumenthal (1996) 47–50.

Philoponus gives three reasons why people have doubted the authenticity of the *Categories*, first that other Aristotles existed, second, that other *Categories* written by his associates were said to exist, and third that Ptolemy Philadelphus' willingness to buy books for the library at Alexandria encouraged forgery.[21] Hence there were 40 books of *Analytics* in the Greek library and two of *Categories* of which four and one respectively were judged to be genuine.

71F Olympiodorus, *On Aristotle's* Categories, Prologue (*CAG* vol. 12.1 p. 13.23–35 and 24.12–14 Busse)

[21] C.W. Müller, 'Die neuplatonischen Aristoteleskommentaren über die Ursachen der Pseudepigraphie' *Rh. Mus.* n.s. 112 (1969) 120–6 has a detailed survey of the theories about falsification including all our sources, to which he adds Galen *In Hippocratis* De natura hominum 15 105 (Kühn) *CMG 5* 9,1,55, where the rulers of Alexandria are blamed for offering money for forgeries. See on Müller's views I. Hadot (1990) 152–5.

These two passages are parts of a longer one (13.7–24.20) in which Olympiodorus is considering the genuineness of the *Categories*: there are three reasons for calling this work spurious, the ambition of kings, the goodwill of pupils, and the similarity of names, those of authors, or of works, or of commentaries. The kings included Jobates for the Pythagoreans, Ptolemy Philadelphus for Aristotle, and Peisistratus for Homer. This led to people either writing (spurious) works or getting hold of an old book and giving it a false name. The similarity of names included, for Aristotle, also Aristotle Mythos[22] and Aristotle Paidotribes, of works the similarly named works of Theophrastus and Eudemus, *hupomnemata* (memoranda)[23] of Theophrastus on his own *Categories* and, if and we keep our translation, a commentary of Alexander on Aristotle and Theophrastus together.[24]

p. 24.12–20: Of the reasons for supposing that it is genuine the third (our lines 14–16) is that Aristotle's pupils themselves wrote *Categories*, and the fourth that there were found in the libraries after his death 42 (μ' καὶ β') *Analytics* and two *Categories*.

Elias, *On Aristotle's* Categories introduction (*CAG* vol. 18.1 128.15–18, 133.9–14, and *On Porphyry's* Isagoge, Introduction (*CAG* vol. 18.1 36.32–6, 38.15–16) with Anonymous, *Commentary on Aristotle's* Categories, Introduction (cod. Urbin. 35 = p. 33b25–8 Brandis)

Following the standard procedure, which dates back to Ammonius, Elias tackles the question of the title of the *Categories*, saying that there have been six titles. But Elias' text is faulty, and can be improved from the anonymous commentary of Urbinas 35 given by Brandis. This states clearly that some of Aristotle's colleagues called their works *On the Categories*, others *On the Ten Genera*. At 33b31–3 it refers to Adrastus as the source for information about the books in the library. Was Adrastus the source of all of this?[25]

Paris. suppl. Gr. 678 has part of Elias on the *PA*. It says that Elias was a prefect before becoming a philosopher. But Westerink thinks he became a prefect when quite old. See Wildberg (1990) 38–9, 42–5.

Ilsetraut Hadot in Blumenthal and Robinson (1991) 176 n. 7 wants to restore the commentary on the *Categories* to David.[26]

---

[22] This was a disciple of Aeschines the Socratic, named by DL (5 35) as one of the other Aristotles; the paidotribe features in the same list as having been mentioned by Aristoxenus in his life of Plato.

[23] The meaning of *hupomnêma* is uncertain, and it has been rendered variously as "notebook", "commentary", "memorandum" and "dissertation". It is normal for the Aristotelian commentaries to have the word *hupomnêma* (singular) in their title. David Blank (AmmonBlank 2–3) is helpful. See also Blumenthal (1996) 21,54–5, and above p. 11 on **1** line 237.

[24] Barnes regards this suggestion as anomalous, and wants in lines 12–13 to translate "actually" instead of "entirely" and "he has written" instead of "was written by him".

[25] The Anonymous says correctly that Adrastus called the work Πρὸ τῶν τόπων ("*Preliminaries to the* Topics") and Archytas called it Περὶ τῶν καθόλου λόγων ("*On Universal Terms*"). Elias has transposed these two titles.

[26] But her references to her *Simplicius* in note 7 are mainly unhelpful. Only at 167–8 does she give arguments.

71G Anonymous Coislinianus, *On Aristotle's* De interpretatione (*CAG* vol. 4.5 p. xxiii.29–30 and 34–6 Busse)

This work is found in several MSS including Coislinianus 160 and Laurent. 72,1. The focus of attention is the genuineness of the *DI*, but a similar tale to that on the *Categories* is told about the proliferation of works in emulation of Aristotle. It is a puzzle that the *On Style* should have been related to the *Categories*, since we know enough about Theophrastus' *On Style* to see that it was nothing like the *Categories.* See the passages from Dionysius of Halicarnassus at **688** and **692**. Eudemus, however, wrote an *On Style* which was connected with logical matters.[27] On the other hand the writer was correct in relating *On Affirmation and Denial* to the *DI*. See on **72A**.

Busse says this work is too feeble to be by Philoponus; it is by a Christian younger than Philoponus. He (p. xx) derives it from Ammonius, "but the sentences with which we are concerned have no equivalent in his" (Ammonius') "extant commentary on the *DI* and seem to have been made up by the scholiast using material derived from other commentators" (Gottschalk 247). Gottschalk regards it as worthless.

71H Abū-l-Farağ Ibn-aṭ-Ṭayyib, *On Aristotle's Categories*, Introduction (ed. by ʿA. Badawī, in Mohaghegh & Izutsu (1974) 82 [European numerals], and in Ferrari, as listed below, 29)

Abū-l-Farağ's commentary on the *Categories* has survived in a unique copy in the Cairo MS Ḥikma 1M and was recently edited and studied by Cleophea Ferrari, "*Der Kategorienkommentar von Abū-l-Farağ ʿAbdallāh Ibn aṭ-Ṭayyib* [Aristoteles Semitico-Latinus 19], Brill: Leiden 2006.

This passage is not about the *Categories* by Theophrastus but about those by Aristotle and the question of the authenticity of the latter. This question was discussed, in the commentaries on the *Categories* from late antiquity upon which Ibn-aṭ-Ṭayyib directly drew, in the introductions to the commentary proper, together with other questions (such as the

[27] See Fortenbaugh (2002) 59–83. Gottschalk, 'Eudemus and the Peripatos' also in Bodnar and Fortenbaugh (2002) 31 n. 12 withdraws a previous suggestion that this *On Style* was the work of Eudemus and prefers to connect it with the Περὶ τὴν λέξιν ("*On Language*?") discussed below in connection with Simplicius.

usefulness of the work, its divisions, its style, etc.). One of the arguments in favor of the authenticity of the *Categories* was that "the most eminent of Aristotle's own colleagues accepted the book as genuine".[28] Simplicius, from whom this quotation is taken, appears to be the only extant late antique commentator actually to use this argument. He does not, however, specify the names of these colleagues. The mention of both Theophrastus and Eudemus by Ibn-aṭ-Ṭayyib in this connection would indicate that these two names did appear in his source; it cannot be maintained that Ibn-aṭ-Ṭayyib himself added them although he was quite aware that Theophrastus and Eudemus were indeed among the most intimate of Aristotle's companions. Ibn-aṭ-Ṭayyib knew almost nothing of Eudemus's works,[29] while in the writings of Theophrastus to which he had access in Arabic such a statement does not appear to be present. It can be assumed that Ibn-aṭ-Ṭayyib's source was a commentary subsequent to that of Simplicius in which the implied "colleagues" of Aristotle were actually named. (DG)

1I Ibn-an-Nadīm, the *Index* 7.1, chap. on Aristotle (p. 248.20–1 Flügel)

Müller (1872) 25–26. Peters (1968) 7ff. Gutas (1985) 81.

Ibn-an-Nadīm mentions twice in his *Index* (*al-Fihrist*) a commentary by Theophrastus on the *Categories*, once in his entry on Theophrastus himself (F 252.10–11= **3A**) and once in the entry on Aristotle's) *Categories*, the present fragment. In the former entry he says that such a commentary is falsely attributed to Theophrastus, without naming the source of his information. In this passage, he simply lists Theophrastus along with a number of other philosophers who were supposed to have written a commentary on the *Categories*. He says, to quote him in full:

[28] Simplicius, *On Aristotle's* Categories, *Prooemium*, 18.13 Kalbfleisch: καὶ οἱ σπουδαιότατοι τῶν ἑταίρων αὐτοῦ ὡς γνήσιον ἀπεδέξαντο τὸ βιβλίον.

[29] For all the material on Eudemus available in Arabic see D. Gutas, 'Eudemus in the Arabic Tradition,' in Bodnár and Fortenbaugh, 2002, 1–23; repr. in Gutas (2000) no. VIII.

> *Categories*, translated by Ḥunayn ibn-Isḥāq. Among those who interpreted and commented on it are the following: Porphyry, Stephanus of Alexandria, Allīnus,[30] John Philoponus, Ammonius, Themistius, Theophrastus, Simplicius; <there is also a commentary> by a man known as Theon, in Syriac and Arabic.[31] What is found of the commentary by Simplicius goes up to the [discussion of the category of the] relative.[32] Among the rare commentaries there is a fragment which is attributed to Iamblichus; the Master Abū-Zakariyā [Yaḥyā ibn ʿAdī] said, "It is almost certain that this is falsely attributed to Iamblichus, because I found, interspersed in the text, [the statement,] "Alexander said".' (F 248.20–24).

It is not clear whence Ibn-an-Nadīm derives this list. The initial assumption is that he compiled it himself on the basis of *Categories* commentaries he had either seen or learned about from reliable sources, such as his colleague, the philosopher Yaḥyā ibn-Adī. This would appear to be the more likely alternative. It is more difficult to assume that Ibn-an-Nadīm is here reproducing uncritically a complete list he found somewhere: the reference to "a man known as Theon" would tend to indicate his own doubts about the identity of this person rather than those of his source; besides, the order in which these names are listed does not follow any identifiable rationale, which one would have expected to be the case in a list prepared by a (Greek?) scholar in the subject. It would thus appear that Ibn-an-Nadīm compiled the list himself, most likely on the basis of information he received from Yaḥyā ibn-ʾAdī. The likelihood of Yaḥyā's intermediacy in this regard is enhanced by the fact that Ibn-an-Nadīm cites him right after this passage in connection with the alleged commentary by Iamblichus. Yaḥyā, in his turn, would have derived his information about a commentary on the *Categories* by Theophrastus from the Greek commentatorial tradition that is listed at **71**. (DG)

This is not the place to go over all the arguments that have been put forward about whether Theophrastus wrote a *Categories.* Gottschalk argues that since there is no trace of such a work in the many commen-

[30] This philosopher, who appears a number of times in the Arabic sources as an Aristotelian commentator, has not been identified. The available information on him has been presented in detail by Elamrani-Jamal in DPA I 151–152.

[31] Reading, in accordance with the necessary addition by Müller (1872) 6, *wa-li-rağulin yuʿrafu bi-Ṯāwun <šarḥun> suryāniyyun wa-ʿarabiyyun.* Müller suggests, rightly I believe, that this phrase about Theon is an interpolation at some stage of the recension of the *Fihrist* that has been added in the wrong place insofar as it interrupts the reports about the commentary by Simplicius which precede and follow this phrase.

[32] Reading, following the edition of the *Fihrist* by Riḍā Tağaddud 309,5–6: *wa-yuṣābu min tafsīr Sinbilīqūs ilā l-muḍāf.*

tators who wrote on the *Categories*, it is most unlikely that it should have become known to the later commentators.[33] Frede, on the other hand, lists the evidence in detail and appears more positive. It is possible that originally the *Categories* was not classed as a logical work, and so escaped the net. Further, if, as some think, like Frede who (12–13) suggests that the original title was Πρὸ τῶν τόπων ("*Before the Topics*"), Aristotle's work was not originally called *Categories*, and if Theophrastus took it as a starting-point, he would have called it something quite different. See F.A.J. de Haas, 'Did Plotinus and Porphyry disagree about Aristotle's *Categories*?' *Phronesis* 46 (2001) 492–526. At 510–13 he discusses the meaning of *categoria*. (PMH)

At this point mention should be made of a puzzle in Simplicius *On Aristotle's* Categories 65.4–10 trans. SimpCat (Chase) 80, where, after a long account of problems raised about parts of speech and categories Simplicius refers to three works of Aristotle, the *Methodics*, which is mentioned by Aristotle himself at *Rhetoric* 1.2 1356b20, and which Moraux, *Les listes anciennes des Ouvrages d'Aristote* Louvain: éditions universitaires de Louvain (1951) 66, thinks is the *Topics, the Divisions*, and a *hupomnema* which if not by Aristotle is, he says, certainly by a member of his school. It is called Περὶ (or Παρὰ)[34] τὴν λέξιν ("*On Language*" or, with Chase, "*On Lexical Matters*"). In all three Aristotle is said to have added to the categories "also the *ptoseis*".[35]

One unanswerable question is whether Simplicius himself knew all three works and found them saying the same thing. A second, more promising, question is what the exact title of the third work was. The MSS of Simplicius give both Παρὰ and Περὶ. There is a temptation to relate this work to Galen's *On* (περὶ) *Sophisms based on* (παρὰ) *Language*,

[33] Sten Ebbesen, 'Boethius as an Aristotelian commentator', in Sorabji (1990) 387 reports that Boethius *On Aristotle's* Categories 160A–B says that he will write another work which will give the opinions of different people on what the *Categories* is about. In MS Thott 168 2° fols. 2–3 (Royal Library of Copenhagen) a scholiast refers to "the opinions of different people: 'Aspasii, Alexandri, Theophrasti, Aristotelis, Andronici." But Ebbesen tells me that this scholiast is of no value.

[34] Hayduck prints περὶ with Marcianus 224 (c xi), but Marcianus 225 (c. xiv) and the Venice edition have παρὰ. See now Fortenbaugh (2002) 78–9.

[35] I.e. the cases or inflexions of a word. A translation of both the Simplicius passage and the Dexippus one mentioned below is given by Ross under "Dissertations" in his *Select Fragments* (*The Works of Aristotle translated into English* vol. 12 Oxford: Oxford University Press 1952 106), but he is misleading when he introduces the term "fallacies" into the title, where the Greek has only παρὰ τὴν λέξιν ("based on expression"). See now also Chase's translation.

which is given in the list of his works connected with Aristotle at **73**. In Simplicius παρὰ looks the more probable reading, but κατὰ was the MS reading, changed to παρὰ by the editor, Mueller, who compared Galen's surviving work of that title. That is connected with Aristotle's *Sophistici Elenchi*, and does not refer to categories at all, so that there may be no connection with the one known to Simplicius.

We cannot be sure of the title of that work, but we can accept that there was a work which dealt with *ptoseis*, and it might have been by Theophrastus. Simplicius quotes a few of the actual words about *ptoseis*, but then expands in what appear to be his own words, referring to negations, privations and indefinite terms. We might try relating it to another work known to Simplicius, for which see **666** no. 17b and **683** and Fortenbaugh's commentary. There a work called *On the Elements of Speech* is said to be by Theophrastus and οἱ περὶ αὐτὸν—which may mean anything or nothing—and it deals with various aspects of speech, though *ptoseis* are not mentioned.[36]

A similar and earlier account is given by Dexippus, *On Aristotle's* Categories 33.10–21, trans. John Dillon London: Duckworth 1990 (*ACOA*) 67. Dexippus was a follower of Iamblichus and so of the early fourth century, but his account is less full and cannot be Simplicius' (only) source. All this amounts to no more than faint evidence that Theophrastus wrote something connected with categories, but not one called *Categories*.

72A–B are concerned with the *DI*: here we are on much firmer ground. There is adequate evidence that Theophrastus' *On Affirmation and Denial* is related to Aristotle's *DI*. There is however some doubt about the title.

H. Maier (***A.G.Ph.*** 13 1900 **51ff.**) has argued that the fact that there is no trace of the new terminology used by Theophrastus in the *DI* is evidence for the genuineness of the latter.

72A Boethius *On Aristotle's* De interpretatione, Second edition, 1 Introduction (*BT* p. 11.30–2 and 12.3–16 Meiser)

De Rijk (1964). K. Bärthlein, 'Nochmals Über das isolierte Aussagenwort' *Rh. Mus.* 127 (1984) 253–8.

[36] As noted above at the end of **68**, Schenkeveld has argued that **683** should be counted as logical.

The details of Boethius' life and the dates of his writings are still disputed, but it is agreed that he set out to produce Latin translations of the works of Aristotle and Plato.[37] He succeeded only in giving us a number of works of logic, mainly from Aristotle's Organon but including Porphyry's *Isagoge* or *Introduction*, which had an immense effect on Western medieval philosophy For a recent bibliography see Mignucci (1996) 304 n. 53.

This passage is from Boethius' second commentary on the *DI*.

In arguing about the authenticity of Aristotle's *DI*, Boethius cites a lost work of Alexander, presumably his commentary on the *DI*, and connects Theophrastus '*On Affirmation and Denial* with the *DI*; this is supported by the fact that we have several excerpts from Theophrastus' work which are clearly related to that of Aristotle, and that at **73** Galen refers to his own commentaries on it in a list of several works connected with Aristotle's writings. See also on **71G**. Alexander tended to shorten the title, but it was clear what he meant. The work, whatever its title, was connected with Aristotle's *DI*, and must have discussed denial as well as affirmation.

H. Usener, *Analecta Theophrastea* Diss.Bonn 1858 62 = *Kleine Schriften* I 50–88 Leipzig-Berlin: Teubner 1912, thinks that the original title was Περὶ ἀποφάνσεως ("*On Statement*"), but that seems unnecessary. There is indeed something wrong with the entry in DL's list, which has the book number alpha between the *katapha*seos ("affirmation") and the *kai* ("and"), but in view of the evidence from other writers it seems safe to move the alpha to the end of the entry. There is also the title Περὶ ἀποφάνσεως ("*On denial*") in the second alphabetical list given by DL, and it is this that Usener wanted to alter. Otherwise we are left with the title *On Denial*, which is not found anywhere else. Such a work could however have existed.

Bärthlein, in an appendix to his main article, surveys the problem of the authenticity of the *DI*, and gives much weight to Theophrastus as evidence in favour of that.

The question of what was originally in DL's list is a different one. Something has gone wrong somewhere. As so often with this kind of material, I do not think we can get beyond interesting conjectures. Boethius' remarks about Theophrastus' methods have been pressed because he is the only one to go into detail about these methods. Since

[37] See *On Aristotle's* DI[2] 79–80.

he is writing about the *DI* it is fair to assume that what he says is correct for that work and that in it Theophrastus did spread himself on matters that Aristotle had left relatively untouched. Whether it applies to all other works relating to Aristotle's is less certain, but it does apply to the *Physics* and the *On the Soul.* Priscian of Lydia, in his *Metaphrase of Theophrastus on Sense-perception* refers in two places (7.22 and 36.6) to Theophrastus' methods in similar terms to those of Boethius.

Courcelle (1969) 282–3 says that Porphyry was an important source for Boethius, and even suggests that he was the only source. But here Alexander looks like the source; Tarán (AnonTarán vii–viii n. 10) considers and rejects Courcelle's view that Boethius is closely related to Ammonius.[38] At 7.5–7 Boethius does say that he uses Porphyry most, but at **10** he brings in Alexander's *DI* commentary and begins the discussion of species/genus. See also on **80**. Shiel (1984) thinks Boethius did not *use* Ammonius.

It has been suggested that the subject of the next sentence (16–18) *addit quoque hanc causam, quoniam Aristoteles quidem de syllogismis scribere animatus numquam id recte facere potuisset, nisi quaedam de propositionibus adnotaret* is Theophrastus. But this is surely an extra reason given by Alexander for the genuineness of the *DI.*

72B Ibn-an-Nadīm, the *Index* 7.1, chap. on Aristotle (p. 249.1–3 Flügel)

Müller (1872) 26. Peters (1968) 12. Gutas (1985) 81.

The context in which Ibn-an-Nadīm mentions Theophrastus here is in the chapter on Aristotle, in the discussion of the individual books. He says:

> *De Interpretatione*. The text was translated by Ḥunayn into Syriac and by Isḥāq into Arabic. Commentators: Alexander, [commentary] not available; John Philoponus; Iamblichus; Porphyry; the epitome of Stephanus; by Galen there is a commentary, rare and not available; Quwayrī, Mattā Abū-Bišr, al-Fārābī. And by Theophrastus <***>. Some of the abridgments: Ḥunayn, Isḥāq, Ibn-al-Muqaffaʿ, al-Kindī, Ibn-Bahrīz, Thābit ibn-Qurra, Aḥmad ibn-aṭ-Ṭayyib, ar-Rāzī.

[38] Mignucci (1996) 283 n. 12 gives a very full bibliography for this point. But it is not clear whether he accepts Shiel's view that Boethius used scholia in a Greek text as the basis for his commentaries. Monika Asztalos, 'Boethius as a Transmitter of Greek Logic', *Harvard Studies in Classical Philosophy* 95 (1993) 367–407 also doubts the scholia theory. Magee (see fn. 236) XLI–XLII treats it with respect but does not accept it.

The text is clearly lacunose here. "And by Theophrastus" (*wa-li-Ṯāwufrasṭus*) introduces a sentence that has not been completed; it is not a continuation of the previous sentence. Ibn-an-Nadīm frequently uses this expression, i.e., the preposition *li-* ("by") followed by the name of the author, to add an item to his enumeration and to say something specific about it, not merely to list the name of an author in a long list with other names: see the sentence about Galen that just precedes this passage, and further cf., e.g., p. 249.23, *wa-li-l-Fārābī tafsīr hāḏā l-kitāb wa-lahū muḫtaṣar fīhi*, etc. Furthermore, it is clear that the name of Theophrastus does not belong to the previous list of names because it is out of place: In the list under the rubric "Commentators," Ibn-an-Nadīm first lists, in apparently what was intended as a chronological order, the Greek scholars who wrote commentaries and ends this part of the list with a special entry on Galen, remarking that his commentary, rare to begin with, is no longer available. Then follows his list, again in chronological order, of scholars writing in Arabic. At the very end of this section on commentaries, and before starting the new rubric on "Abridgments," Ibn-an-Nadīm intended to say something special about Theophrastus: a work on *De Interpretatione* which could not be listed along with the others as a commentary by the mere citation of Theophrastus's name because of a distinctive feature. We do not know what he intended to say, but I would guess that he meant to note that the book of Theophrastus on the same subject as Aristotle's *De Interpretatione* had a different title, *On Affirmative and Negative Propositions*, a piece of information Ibn-an-Nadīm or his source could have derived from al-Fārābī's statement to this effect: see Gutas (1985) 81, no. 2.1 and below at **81B**. The reason for the existence of a lacuna here is also a matter of speculation, though it would appear most likely that the space was initially left blank by Ibn-an-Nadīm for the purpose of checking the reference and that it was never filled in.[39] This happens frequently in the *Fihrist*, most commonly with death dates. (DG)

[39] This would appear to be corroborated by the fact that Ibn-al-Qifṭī, who copies Ibn-an-Nadīm for his entry on the *De Interpretatione*, also had a lacunose copy of the *Fihrist*: the phrase *li-Ṯāwufrasṭus* appears without any complements, except that in some manuscripts whose variants are listed by the editor, Lippert, the preposition *li-* appears as an *alif*, thus taken by readers (and perhaps by Ibn-al-Qifṭī himself) as part of the Greek name. (DG)

73 Galen, *On His Own Books* 14 (*BT* p. 122.19–123.9 Mueller)

Galen found that books were attributed to him which were not his, and decided to list his genuine works, though in fact some genuine works are omitted.[40] The text we have is not good, but we can improve pp. 122–3 from 118, 119, and 121. In our passage Galen lists under "On the works relating to Aristotle's Philosophy" first commentaries on the logical works, indeed on the whole of what came to be called the Organon[41] with the exception of the *Topics*, then six commentaries on Theophrastus' *On Affirmation and Denial*, which alone are ascribed to him by name. But the next title, *On How Many* (*Senses*), is supported by **133** and **134**. The next item, *On the First Mover Unmoved*, without attribution, raises difficulties. Galen overwhelmingly uses *Eis* ("On") with the titles of works, as opposed to *Peri* ("Concerning") followed by subjects. We should probably therefore take the *First Mover (is) Unmoved* as a title, and the work of Galen would be a discussion of a subject from Aristotle which, in this position, might have been discussed by either Theophrastus or Eudemus.[42] (See on Metaphysics FHS&G **247**). There follows something on Eudemus' *On Style* (or *On Expression*), and then four more or less logical works of which the middle two are *On Problematic Premises*[43] *and Syllogisms* and *On Syllogisms with Mixed Premises*; these are on the subjects on which Theophrastus and Eudemus collaborated (See **102, 103** and **105–8**): could Galen have been writing about works of joint authorship? See on **106A**. (The other two are *On Proofs connected with Why* and *On Sophisms based on Language*, a work which is extant[44] (vol. 14 pp. 582–98 Kühn). This may contain a quotation from Eudemus (29W), but Fortenbaugh (2002) 70–5 has serious doubts about that.

[40] Jaap Mansfeld, *Prolegomena—Questions to be Settled Before the Study of an Author, or a Text* Leiden: Brill 1994 Chap. IV "Galen's Autobibliographies and Hippocratic Commentaries' is useful background. V. Boudon, "Galen's *On my own Books*: new material from Meshed, Rida, tibb.5223' in Nutton (2002) 9–18 uses a new Arabic MS to extend our knowledge of the text, but only on medical matters.

[41] See on **72B**.

[42] G. Strohmaier, "The Uses of Galen in Arabic Literature". Nutton (2002) 113 takes from al-Biruni's *India* the title of a work by Galen which is given in the Latin translation as *In primum movens immotum* (*On the First Mover Unmoved*).

[43] At **100C** we have an excerpt from the Arabic translation of Alexander's work on Galen's *Essay on the Possible.* How that is related to other works on logic we do not know.

[44] Translated in Edlow 1977, but see the review in *Journal of Hellenic Studies* 98 (1978) 180–1. The text has been edited by Sten Ebbesen (1981).

In lines 3–4 some supplementation is required, but Mueller's addition, taken from p. 118, is surely correct.

At the end of line 6 *ta* ("the") is exceptional. Sharples asks if we should take the meaning as "the commentaries are three", but why should this form occur only here?

Sharples points out that only with the categories and the unmoved mover is the word *Peri* (About) missing.

For the last title, *On Sophisms based on Language*, compare what is written after **71I** about Simplicius, *On the Categories* 65.4–10, on Τὰ περὶ τὴν λέξιν ("*Things connected with expression*").

Galen, *On his Own Books* 11 (*BT* p. 118.14–16 Mueller)

Galen says that none of his commentaries on Aristotle's logical works was written for publication, nor were the six he wrote on Theophrastus' *On Affirmation and Denial*.

-6 are connected with theses. All that is clear is that Theophrastus wrote a large number of them, and that they were attractively written. See also **135**.

4 Aelius Theon, *Preliminary Exercises* 2 (*RhGr* vol. 2 p. 69.1–4 Spengel)

Aelius Theon was a rhetorician of the first or second century AD. For a discussion of Theophrastus' theses see Fortenbaugh, 'Theophrastus, the *Characters* and Rhetoric' in Fortenbaugh and Mirhady (1994) 23–4 and Fortenbaugh (2003) 224–43. Fortenbaugh thinks that Theophrastus' *On Marriage* (**486**) may have been a thesis. If so, it explains Theon's selecting of Theophrastus' theses as fine examples.

Theon indicates that these theses were valued for their suitability for study by students of rhetoric, listing them among a series of noted works, like the funeral speeches of Plato, Thucydides, Hyperides and Lysias. Only Aristotle and Theophrastus are named as writers of theses. See also on **89** and **135**.

In line 3 the last word should be ἐπιγραφόμενα.

5 Diogenes Laertius, *The Lives of the Philosophers* 4.27

This tells us only that there were theses of Theophrastus known to Crantor, a contemporary of Theophrastus but an Academic, and that they were flowery. David Sedley (1999) 143 translates the penultimate word by "with an oyster-shell", and sees this as a jibe by Crantor.

76 Hippolochus, *Letter to Lynceus*, in Athenaeus, the *Sophists at Dinner* 4.5 130D (*BT* vol. 1 p. 29.5–11 Kaibel)

This tells us less about the nature of a thesis than about the way of life of Theophrastus and his students. For Lynceus see on **18** 10.

In this letter Hippolochus is boasting of the wonderful time he and his friends are having in Macedon, and contrasting their position, eating at a rich man's table and being in the market for houses, fields and slaves, with that of Lynceus and his friends listening to the theses of Theophrastus and eating thyme, rocket and rolls, which are not, for him, delicacies. Compare Athenaeus 2 55 where Menedemus is mocked by Lycophron for dining on lupines.[45]

Logic is not a Part of Philosophy

77 John of Italy, *Various Questions* 16 (p. 18.1–22 Ioannou)

Westerink (1961). P. Hadot, 'La logique, partie ou instrument de la philosophie?' in Hadot, I. (1990) 183–8. For further references see *PCSB* vol. 3 32–6.

John of Italy was born about 1023 in Campania, but in about 1049 he went to Constantinople where he stayed. He was a student of Michael Psellus, and according to Anna Comnena was a very popular professor of philosophy, teaching among other things the dialectical arts of Aristotle. See Thomas Conley, 'Notes on the Byzantine Reception of Peripatetic Rhetoric' in Fortenbaugh and Mirhady (1994) 230–5 and, for his life, Wilson (1983) 153–6.

It is only in this passage, after a long line of earlier witnesses, that we have a named reference to Aristotle and Theophrastus, or rather those around (*hoi peri*) Aristotle and Theophrastus,[46] who are introduced as

[45] But generally the Peripatetics had a reputation for high living. See H.B. Gottschalk, 'Continuity and Change in Aristotelianism' in R. Sorabji (ed.), *Aristotle and After* London: *BICS* Suppl. 68 1997 110.

[46] The combination of Aristotle and Theophrastus with "*hoi peri*—the associates of" is unusual. It has become a matter of received wisdom that *hoi peri* X expressions amount only to X, but that cannot always be the case. M. Dubuisson, Oi amphi tina oi peri tina Evolution des senses et des emplois. Diss. Univ. de Liège (1980) says that in the Hellenistic period *hoi peri* X can mean X but need not, and in the Byzantine period

holding a view about the relations between philosophy and logic which is indeed a Peripatetic view but not to be found in Aristotle and probably not to be attributed to Theophrastus either. The many other writers who dealt with this topic refer only to the Peripatetics.

The earliest writer known to us to have dealt with the problem of whether logic is a tool or a part of philosophy is Alexander of Aphrodisias, writing about 200 AD. He opens his commentary on the *PA* with this topic. Much of what he says is remote from John of Italy, but he does write about a problem already in existence.[47] We do not know however when it began, and there are no obvious candidates. At AlexPA 3.2–4 trans. AlexPA (Barnes) 43 Alexander refers to ancients who developed logic as far as was useful and called it a tool. According to DL 7.39 Zeno himself treated logic as one of the three parts of philosophy,[48] so that the question could have arisen in Theophrastus' lifetime. It is therefore worthwhile to explore the arguments, knowing that we are being speculative. Gottschalk (1987) 93 suggests that Andronicus began it.

Of the relevant texts noted in the apparatus, that of Philoponus gives the best account parallel to John's lines 4–8. In a very long section (*On Aristotle's* PA 6.19–7.23) he shows that what is incomplete in the Stoics' argument is that they do not have ἢ ὄργανον ("or a tool") in their argument, but deal only with the notions of *meros* and *morion*, and conclude that logic is not a *morion* of philosophy, and is therefore a *meros*.[49] The

it depends on the model followed. Here, in this very late author, and in this context, we need not worry too much about what was intended. Alexander, *On Aristotle's* PA 1.3–3.4 (trans. AlexPA (Barnes) 41–3) distinguished his two groups by *hoi men* and *hoi de*, but at 3.3 refers to the *archaioi*. These should be Peripatetics after Theophrastus and Eudemus. At 284.12–13 he says that *hoi peri Aristotele* ("those around Aristotle"), being concerned with the usefulness of logic, gave up certain argument forms, whereas "those from the Stoa" developed useless forms. "Those around Aristotle" can hardly here be Aristotle, but who they were we cannot say. See also Bocheński 65 n. 223. Galen also referred to *hoi peri Aristotelen* at 16 1 on relational syllogisms, for which see after **97**.

[47] Plotinus (205–70 AD) 1 3.5 9–10 says that dialectic must not be thought of as a tool of the philosopher. He is here dealing with Plato's dialectic, but clearly thinks of logic as a part of it. See Westerink 131–2 about the place in a course of the question of the function of logic. A.C. Lloyd, 'Neoplatonic Logic and Aristotelian Logic II', *Phronesis* 1 (1956) 151 says of Porphyry's time that the view that logic was a tool of philosophy, not a part, was a survival from "an out of date vendetta with the Stoa", and that the "Neoplatonists tended to claim that it was both."

[48] See K. Ierodiakonou, 'The Stoic Division of Philosophy' *Phronesis* 38 (1993) 57–74 for a study of the Stoic account.

[49] This point is also made by Schol. Brandis 140b4–7. He also sets out the arguments of the Academics at some length.

Stoics have a clear distinction between a *meros* and a *morion,* for which LSJ 2 p. 165 quote Ammonius *On Aristotle's* PA 8.34–5 to the effect that *meros* is a big *meros* and a *morion* is a *meros* and a *meros* of a *meros*; Philoponus gives a different account. This suggests that there was some unclarity about the meanings of *meros* and *morion*; presumably these terms were used by the Stoics and taken over by the Peripatetics.[50] Olympiodorus, *Prolegomena to the* Categories, which is a student's account of Olympiodorus' lectures, has a lecture (14.12–18.2) on whether logic is a part of philosophy, which is both formal and full. Busse notes that this has no parallel in Ammonius and Philoponus, though the latter does have part of it, but it does have a parallel in Elias and David. At Olympiodorus 16.10–30 there is a parallel to John's first argument.

The other points made by John are to be found in various places in the commentators. Olympiodorus, 16.10–30 makes the point of our line 9 that philosophy and logic have different matters and different goals; that philosophy survives the loss of logic (14–20) is made by Ammonius *On Aristotle's* PA 10.9–11, Olympiodorus 17.3–17, Philoponus 8.27–9, Elias *On Aristotle's* PA 135.30–1 and 137.33–138.1, and the anonymous *De arte logica* p. xi–xii.[51] On our 4–6 the scholium at Brandis 140b7–11 which Westerink (132) thinks is from Stephanus also deals with the Stoic omission of "or an instrument". There is nothing in all of this that is obviously late, but it still seems unlikely that it all goes back to the generation of Aristotle's immediate successors. See T-S Lee (1984) 44–54 for later commentators on this topic. It is worth noting that all terms used here can be mathematical: *aphairesis* = subtraction, *diairesis* division, *prosthesis* addition, while *sunthesis* in this context means the production of a whole which will be destroyed if divided. In LSJ a) it is opposed to *analusis*: b) it can mean addition. But it is also frequently used in a more general sense.

In line 10 there seems to have been a simple printing error in Ioannou's edition, which has *horoi*, whereas in the Index verborum p. 176b10 the correct *holoi* is given. Compare Olympiodorus p. 16.12.

This passage is not in either Graeser or Repici.

---

[50] See also Flannery 17 n. 45.

[51] This is found as a sequel to the *Vita vulgata* of Aristotle.

Statement

Most items in this section are related to *DI*. In almost all Theophrastus is mentioned alone, and in all except the odd **87C** and **87E** just "Theophrastus" and not even "hoi peri Theophraston" is found. It seems probable then that here, as opposed to his work on syllogistic, Theophrastus worked alone, and not with Eudemus.

78 Ammonius, *On Aristotle's* De interpretatione 4 17a1 (*CAG* vol. 4.5 p. 65.31–66.10 Busse) trans. AmmonBlank 72

W.W. Fortenbaugh, 'Theophrastus, Fragment 65 Wimmer: is it important for understanding Peripatetic rhetoric?', *American Journal of Philology* 111 (1990) 168–175 and in Fortenbaugh (2003) 15–21. id. in his Rhetoric Commentary, after **682**. H.B. Gottschalk, 'An Errant Fragment of Theophrastus', *CQ* ns. 42 (1992) 529–33. (For a fuller list see Fortenbaugh (2005) 240–1.)

For Ammonius see on **71A**.

Much of this is on rhetoric rather than logic. The passage on which Ammonius is commenting, *DI* 4 17a1, refers to rhetoric and poetry as being different from logic (17a5–6). Our passage may well come from Theophrastus' *On Style*, and is no. 24 in M. Schmidt, *De Theophrasto rhetore* Halle (1839) 52–3; see **666** 17ab.[52]

Gottschalk suggests that Ammonius or another commentator, faced with very brief remarks by Aristotle and Theophrastus, looked in their rhetorical works and came up with this as the next best thing. But Boethius (**72A**) says that Theophrastus elaborated on some parts of Aristotle, and he may have expressed ideas like this in his *On Affirmation*. A similar point but about *endoxa* ("reputable statements") is made by Alexander in his *On Aristotle's* Topics 19.22–9: truth depends on the object, but being reputable on the preconceptions of the audience.

Fortenbaugh argues that the point is primarily logical, and that the rhetorical trimmings may be the work of Ammonius, though he thinks that the basic distinction between the factual account and the way in which it is expressed is original.

[52] Fortenbaugh (1990) 168–9 surveys earlier approaches to this as rhetoric. Gottschalk 530–1 covers many earlier studies.

Graeser 59–60 refers to *DI* 4 17a4–7. G.M.A. Grube *The Greek and Roman Critics* London: Methuen 1965 106 refers to Aristotle's *Rhetoric* 3.2 1404a9–12, which is on a similar matter.[53] *Schesis* is rarely used by Theophrastus, but Fortenbaugh n. 11 points out that it does occur in Priscian of Lydia 21.4–8 (**282** 30–4), in what seems to be a genuine quotation from Theophrastus' *On the Soul.*

Gottschalk refers to our **671** where Quintilian says that Aristotle and Theophrastus treated epideictic oratory as solely concerned with listeners.

Anonymous Coislinianus, *On Aristotle's* De Interpretatione (*CAG* 4.5 xxiii.15–17)

This is a condensation of **78**, but it is used in connection with an attempt to explain the meaning of the title *Peri Hermeneias.* It contains nothing of value for our knowledge of Theophrastus.

79 Boethius, *On Aristotle's* De interpretatione Second edition 1 Introduction (*BT* p. 9.24–6 Meiser)

For Boethius see on **72A**.

Boethius gives the full title, *On Affirmation and Denial.* We have used "indicative" here for "*enuntiativa*", but in **80** we have used "statement" for "*enuntiatio*".

80 Boethius, *On Aristotle's* De interpretatione 1 16a1, Second edition 1.1 (*BT* p. 17.24–7 Meiser)

De Rijk (1964)

Boethius is here outlining a disagreement between philosophers about whether *enuntiatio* (*protasis, apophansis*) is equivocal. Porphyry argued that affirmation and denial were species of the genus *enuntiatio*, but Alexander said that *enuntiatio* was predicated equivocally of affirmation and denial, and not as a genus. His argument was that affirmation is prior to denial, and there cannot be a genus of prior and posterior. Boethius then gives our **80** and goes on to say that here (*hic*) Porphyry refutes the argument of Alexander, saying that we must not think that everything which is prior to something else in any way whatsoever cannot be placed under the same genus, but only those which are prior and posterior in their essence and substance.

[53] Bocheński (39) tried to relate it to *DI* 1 16a3, but that is on a different matter.

From all this we can get with certainty only that Porphyry wrote commentaries on Theophrastus' logic in which it was appropriate to discuss this matter.[54] Since we do not have the relevant commentary of Alexander we must be content with **81A**. But it is reasonable to suppose that Alexander discussed the matter at greater length in his *De Interpretatione* commentary, and found Theophrastus of considerable interest. Ammonius, *On Aristotle's* DI (*CAG* 4.5 15.22–9 Busse trans. AmmonBlank 24) agrees with Boethius and declares his support for Porphyry.

Bocheński (40–41) argues that Alexander's view that *protasis* is not the genus of affirmation and negation comes from Theophrastus.

There has been much argument about Boethius' sources, and some think that all this comes from Porphyry only. See on **72A**. In this particular case the matter is unimportant, for we get the essentials from Alexander's own account in **81A**.

A Alexander of Aphrodisias, *On Aristotle's* Prior Analytics 1.1 24a16 (*CAG* vol. 2.1 p. 11.13–16 Wallies) trans. AlexPA (Barnes) 56.

This passage comes at the end of a long discussion of the *protasis* (10.10–11.14), in which Alexander distinguishes various accounts of "*protasis*" in Aristotle. It is in this context that Alexander says that Theophrastus defined "affirmation" and "denial", but not "*protasis*", and concludes that he thought that "*protasis*" has several senses. This may be no more than a deduction from the absence of a definition of "*protasis*", but if Bocheński, 40–1, is correct there was more to it than that, and Theophrastus would be the source of the view Alexander gives above (10.30–11.13) that statement is not the genus of affirmation and denial.

B al-Fārābī, *On Aristotle's* De Interpretatione, Introduction (p. 22.1–9 Kutsch and Marrow)

Zimmermann (1981) 6b, lxxxiv–lxxxv. Gutas (1985) 81.

Following the Greek tradition, al-Fārābī makes the point in this passage that the *De Interpretatione* is not about premises, which are propositions

[54] Even this is doubted by Courcelle (1969) 283, who thinks it might refer only to the discussion of Theophrastus which Porphyry included in his commentary on Aristotle's *DI*.

forming a syllogism, but rather about questions of the type "p or not-p?", the "things sought" he mentions, *al-maṭlūbāt*, a term used to translate the Greek *zētoumenon*. This question contains an opposite pair, and Aristotle's work, he claims, seeks to discriminate this opposite pair from all others. The title of Theophrastus' work, *On Affirmative and Negative* [*Propositions*], is brought here as a corroboration of this point. See the analysis and discussion of this point in Zimmermann (1981) 3 and note 3, and lii–liii. (DG)

82A–E Brunschwig (1982). Mignucci (1998).

All but **82A** are on *DI* 7. The subject is the indefinite, indeterminate or unquantified, to which Theophrastus seems to have paid particular attention. AlexPA (Barnes) 131 n. 128 wants to separate **82A** from the others, on the ground that they deal with a different topic, but both seem to be about the *On Affirmation*. See on **82A**. Mignucci 41 sees here evidence that Theophrastus was further thinking about the nature of the quantifiers introduced by Aristotle. We might see him as a name-giver, but it is disturbing that three Greek writers use three different words for "indefinite"—*adioristos* (Alexander and Tarán's Anonymous) *aprosdioristos* (Tarán's Anonymous), *aoristos* (Ammonius). We can distinguish various aspects of the situation:

1) "Socrates is just"—definite proposition
2) "Some man is just" (**82BE**)—indefinite[55] particular proposition
3) "Some man walks" and "Not every man walks" (**82CD**)—indefinite particular propositions
4) "Phainias has some knowledge" (**84**)—indefinite proposition, which may be true along with its contradictory.
5) Aristotle in *DI* uses *aoristos* of what Theophrastus calls *ek metatheseos* (**87**).
6) At **92** Theophrastus is said to have dealt with syllogisms with some indefinite premises.
7) At *PA* 1.13 32b10–11 Aristotle distinguishes a sense of *endechesthai* (be possible) as being *aoriston*.

[55] In the translation we have used "unquantified", but that does not suit what we have here. And note that it is not of *tis* ("some") that the notion of being indefinite is used here but of propositions. *Prosdiorismos* is not in Aristotle, though *prosdiorizein* is.

8) At *PA* 1.5 27b20,28 Aristotle uses "*ek tou adioristou*" ("from the ambiguity of the particular negative proposition"—Ross). Cf. 1.4 26b14–20.[56]
9) In prosleptic propositions it is said that the third or middle term is indefinite (*aoriston*) (**110**).

82A Alexander of Aphrodisias, *On Aristotle's* Prior Analytics 1.4 26b15 (*CAG* vol. 2.1 p. 66.2–10 Wallies) trans. AlexPA (Barnes) 130–1.

What Aristotle is discussing is: "to some C the B does not belong" which, he says, is true both if B belongs to no C and if it does not belong to all C because it does not belong to some C.

If Alexander refers here only to the *On Affirmation* and not to Theophrastus' *Prior Analytics*, and says only that he mentioned this sense of indefinite in the *On Affirmation*, it seems likely that he said it as an addendum to the discussion of "indefinite" in another sense indicated in **82BCDE**. Alexander relates the word *adioristos* directly to Theophrastus. In the *DI* Aristotle does not use *adioristos*,[57] and when he uses *aoristos* the sense is slightly different. But in Chap. 7 he talks about such statements without using the name.

The passage is puzzling: what is it that Theophrastus is supposed to have added to what Aristotle said? Graeser 64 suggests tentatively that we have here an attempt by Theophrastus to relate *PA* and *DI*: there is no treatment of individuals in the former as there is in the latter, and this might indicate that the latter is a later and more developed work. But it is difficult to be sure of the sense of the end of our passage. What appears to be meant is that there are two senses of "indefinite", the one (*men*) being true when all are true and also when some are *and* some not, the other (*de*) true both when both are true and when one only is true. What is the difference? The first refers to objects in the plural, and the second appears to refer rather to a singular item, which might be a proposition. Can Theophrastus have been distinguishing between two senses of the term as used by Aristotle?

See also on **92**.

Since this is the first appearance of the verb *huparchein*,

[56] At **Appendix 1** is a fragment possibly by Theophrastus and connected with *PA* 1.33 47b29–39 containing the words *protasis adioristos* ("indefinite premise"). *Adioristos* does not appear in that passage of Aristotle. See after **97**.

[57] See the table by Zimmermann in Stern (1972) 530.

I refer here to the discussion of it and its Aristotelian alternatives in Barnes (1996) 184–5. Barnes argues that the formulae concerned are artificial.[58]

82B Boethius, On Aristotle's *De interpretatione* 7 17a38, Second edition 2.7 (*BT* p. 140.3–12 Meiser)

The following four passages are all on *DI* 7. Aristotle is here exploring propositions with universal and with singular subjects, and with negatives and how they are opposed. He does not refer to propositions beginning with "some". Rather, at 17b26–37 he deals with "all" (universal) and "Socrates" (singular), and finally with propositions beginning e.g. with "man", or, as Ackrill 129 urges, "a man". In the *PA* 1.1 he has (24a16) propositions that are universal, particular (*en merei*), or indefinite (*adioristos*). Neither "Socrates" nor "a man" appears here, and "indefinite" refers to propositions like "of contraries there is the same science".[59]

In all our passages we have "some man" (*quidam homo, tis anthropos*); at 8 18a17 "some (*tis*) man is white" forms one of Aristotle's six examples. Theophrastus is said to have called propositions like "some man is just" "indefinite particulars". Note that it is the propositions that are so described. It could be that he is aligning these with the cases given by Aristotle like "of contraries there is the same science".

It looks as if lines 5–8 are also from Theophrastus, but the grammar is puzzling: *quod* should refer back to something, but it itself is neuter and of the only two possibilities *pars* is feminine and *homo* masculine; the sense of the passage requires that it should be *homo*; further why does he say that it is universal in name or in nature? This might conceal a problem about the nature of universals, or possibly a view that some words are just words, while others are universal in a more serious sense. The word "property" (*proprietate*) may look back to lines 2–3, where it is said that individuals differ in their properties; if so one might suppose

[58] Charles Kahn, 'On the Terminology for Copula and Existence' in Stern (1972) 151–4 has a wide-ranging study of the history of *huparchein*.

[59] J.W. Thorp, "Aristotle's use of Categories' *Phronesis* 19 (1974) 253 has a bold suggestion that in a lost work Aristotle did analyse unquantified propositions like "Man walks" as universal propositions with the predicate in the potential mode. So "Man walks" would become "All men are capable of walking". He adduces *PA* 1.28 43b14–15 as evidence that Aristotle was here uncertain about indefinite statements.

that that part too belongs to Theophrastus, but clearly the reference to Cicero cannot be his.

The passage is followed by "*unde universale vocavit...*" in which the subject of "*vocavit*" could grammatically be Theophrastus, but it is clear from what follows that Boethius is continuing his quotations from Aristotle.

Siger de Curtraco, *On Aristotle's* DI 5 17a13 (54.46–8 Verhaak).

Siger (of Courtrai), who died in 1341, was a Belgian who studied in Paris. He uses Boethius but has taken the name of Theophrastus here to cover the definition of "universal" that is in fact that of Aristotle.

82C Anonymous, *On Aristotle's* De interpretatione 7 17a38 (cod. Paris. Gr. 2064 = p. 32.15–33.6 Tarán)

The MS, of the eleventh century, alone contains this work, as well as Stephanus on the *DI* and Ammonius on the *PA*. Tarán says that this commentary is of the late sixth or early seventh century. He rejects the suggestion that it is the work of Olympiodorus or one of his pupils, and despairs of discovering its author. The title it has was added in red ink by a fourteenth or fifteenth century hand. There may have been a title originally, but it is clear that some pages have been lost at the beginning. The work is largely dependent on Ammonius' commentary on Aristotle's *DI*, but also uses another source, possibly Syrianus, who is the latest philosopher named. Tarán thinks the author belonged to the Alexandrian school, and worked either in Alexandria or in Constantinople.

Here the word we translate as "indefinite" is *aprosdioristos*, and it is clearly related to the word *prosdiorismos*, for which we have used "quantifier", although that word has overtones of modern logic which may not always be appropriate.[60] But here the word so described is *tis* ("some") and it is later called "particular quantifier". From this we pass to Theophrastus' term for a proposition like "Some man is walking", namely "unquantified particular", though later in the same commentary it is given as "particular qua unquantified". It may be significant, however, that the only two points about Theophrastus in this work as we have it (in **82CD** and **87E**) are concerned with the names which Theophrastus

[60] At Stobaeus, *Anthology* 4.2.20 (p. 130.1 Hense). Theophrastus in his *On Contracts* (our **650**) uses *prosdioristeon* in the sense of "one should add the following determination" which refers to conditions in which a purchase and sale may be invalidated.

gave to certain expressions. One might suppose the use of a logical handbook from which these items were taken. The word translated "particular" is *merikos*, which is not found in Aristotle; indeed there is no equivalent to it in Aristotle's *DI*, though elsewhere Aristotle has forms like *epi merous* and *kata meros*. Aristotle uses periphrastic expressions like καθόλου…ὄντος τοῦ ἄνθρωπος οὐχ ὡς καθόλου χρῆται τῇ ἀποφάνσει ("Although the [word] man is universal the proposition is not used as universal") (*DI* 7 17b10–11).[61]

In line 4 *phonê* ("word") must refer to *tis.*

Socrates and Plato are mentioned here as in **82B**, but Cicero is not.

In line 5 Mignucci (1998) 41 n. 7 suggests changing *adioriston* to *aoriston*, but perhaps not seriously.

In line 9 "the ancients" are said to call "universal *qua* particular" what Theophrastus called "particular *qua* unquantified". It is not said who these ancients were, but the word used, "*merikos*" for "particular", was not used by Aristotle, which leaves us with a puzzle. At 31.8–18 we are told that the *archaioi* held that in a *protasis* the relation (*schesis*) between subject and predicate is its matter (*hulê*), while they are its form. This too is a late view. But at 2.1–2 we are told that the *archaioi* did not bother to give a name to what Aristotle called *aoriston onoma* ("indefinite name"), and this suggests that the people concerned belonged to an early period. It is possible that the author had no particular people in mind.

82D Anonymous, *On Aristotle's* De interpretatione 7 17b16 (cod. Paris. Gr. 2064 = p. 37.15–21 Tarán)

This is from a later part of the commentary than **82C**, but it is still part of the section to which **82C** belongs. The device of the square of opposition is commonplace, but the addition of the reference to Theophrastus is not. The writer appears to say that the diagram is his own, used to clarify his arguments. We have extended the passage to cover a further sentence, but that may not be based on Theophrastus.

It is probably not significant that the author says: to make things clearer "we will do it in a diagram." He has other diagrams, at 41.15–43.4, 73.9–18, where he says that he does it for the sake of clarity, 77.9–17, and 107.6–15 with a similar remark.

---

[61] Flannery 33 n. 80 finds a similarity between this and Alexander *On Aristotle's* PA 65.26–30 trans. AlexPA (Barnes) 129–30, which he interprets as meaning that an i-proposition is always capable of referring to a plural.

2E Ammonius, *On Aristotle's* De Interpretatione 7 17a38-b12 (*CAG* vol. 4.5 p. 90.12–20 Busse) trans. AmmonBlank 96.

Here the word said to be used by Theophrastus for indefinite is *aoristos*, but the meaning is the same. This word occurs in Aristotle in a logical sense only in *PA* 1.13 32b4–23. See on **103A**.

Brunschwig 368 argues that our lines 8–9 are still attributed to Theophrastus as earlier scholars thought, but more recent editors did not.

In line 2 Barnes wants "state", not "show", and in line 8 "generally" for "simple".

83 Scholium on Philoponus', *On Aristotle's* Prior Analytics 1.1 24a15 (cod. Paris, Gr. 1917 = p. 145a30–7 Brandis)

Brunschwig (1982). Mignucci (1998). Huby (2002) 103.

This MS is of the thirteenth or fourteenth century, but the scholia are said to be of the sixth century. The scholiast is a contemporary of Philoponus, and it is unlikely that he had a source of information unknown to his predecessors.

Aristotle carefully worked out the connections between quantifiers and negatives. I cannot find any passage where he explicitly equates "not to all" with "to some not", but he clearly takes it for granted that it is so, and it is indeed an essential part of his syllogistic, just as ~(x) is equivalent to (Ex)~ in modern quantificational theory. It may be, then, that Theophrastus took up this point. He has been ridiculed by Bocheński (43) and other modern logicians, and several, from Prantl (359,34) onwards, have suggested emendations. They want to rearrange the pairs, so that "not to all" would indicate "at least to one", etc. But this is unnecessary: in English, if we say "not *all*" we normally imply "but most", and when "all" is not emphasized there is no special implication. The choice of one expression rather than the other may indicate that more or fewer items may be involved. All this may be connected with Theophrastus' interest in the linguistic aspects of logical expressions.

Brunschwig 364–5 n. 14 recounts the views of his predecessors.

If the scholiast is correct, and that is doubtful, this passage implies that these criticisms were made in Aristotle's lifetime, and induced him to infer "not to some" from "not to all" or to add "or not to all" to "not belonging to some". It is indeed true that at *PA* 1.4 26a37 he adds a superfluous "or not to all", and at 26b4–5 "or if not to all" to a preceding "not to some".[62]

Brunschwig 364 n. 14 suggests inserting *mê* ("not") after *pleiosin* to get "not to several/many", but Mignucci thinks the scholiast was sloppy. Mignucci, 40 n. 3, supports a suggestion by Van Ophuijsen that *pleiosin* should be rendered "to several", not "to many".

AlexPA (*CAG* vol. 2.1 p. 62.11–13 Wallies) trans. AlexPA (Barnes) 125.

Alexander says that "not to belong to some" and "not to all" both indicate the particular assertoric, differing in expression (only).

84 Scholium on Aristotle's *De Interpretatione* 17b14 (cod. Ambros. Gr.490 (formerly L 93) f.64[v] = *CAG* vol. 4.5 p. xxxiii.34–7 Busse)

Kneales 111–2. Fortenbaugh, W.W. 'Theophrastus, no. 84 FHS&G There's Nothing New Here!', *The Passionate Intellect* ed. Lewis Ayres (RUSCH 7) New Brunswick: Transaction Publishers 1995 161–76 and in Fortenbaugh 2003. Mignucci (1999) 38–9.

This passage has been the subject of widely differing opinions, from the dismissive approach of Busse to the recent enthusiastic interpretation of Mignucci, and the argument of Fortenbaugh that the scholion is misplaced and therefore has been misunderstood.

The MS is of the tenth century, and the scholia are in the same hand as the text of Aristotle.[63] Busse xxxi n. 3 thinks this passage unworthy of Ammonius to whom some later scholia are attributed,[64] and therefore attributes it to the scholiast himself.

---

[62] See also 1.1 24a19 'or not to' and 1.4 26b16 'to some not'.

[63] Fortenbaugh has a detailed study of the text of this and related scholia.

[64] To help others to assess this judgment I add the context of this quotation: the scholiast begins by summarising Aristotle's account, and adds his own views. He takes Aristotle to be saying that "All man is all animal" is false, and goes on to explain that by saying that a horse is not a man. It is at this point that the quotation from Theophrastus comes in, but there is a complication: Fortenbaugh points out that there is a lacuna at the bottom of f.64[r], after which our passage begins at the top of f.64[v]; it is therefore not clear whether we have one or two scholia. Assuming that there was only one, grammatically the scholiast could be responding to his last, mutilated, point and not to Aristotle's more general one. But that is hardly likely, and we must suppose that Theophrastus was responding to Aristotle. Finally there is a reply to Theophrastus' argument, that it is fallacious, for there is one predicate in the affirmation and another in the denial, and this plays upon the homonymy of "knowledge".

We have translated *prosdiorismos* "quantifier",[65] but this may not be what Theophrastus intended. It could mean, as the note indicates, just an extra qualification of any kind. Bocheński (43–6), followed by the Kneales 112, thinks that Theophrastus was attacking what he regards as Aristotle's claim that you cannot quantify the predicate, or even (Kneales) asking whether you can attach a quantifier to a phrase which is neither the subject nor the object of the sentence in which it occurs. Aristotle does indeed say here that you cannot have a proposition like "Every man is every animal." But his remarks are limited to the case where both predicate and subject are quantified. He says nothing about cases where the subject is not quantified, and is not here concerned with pairs of propositions of which one is true and the other false. The same example, "Every man is every animal" is also used at *PA* 1.27 43b21 where the complicated argument is concerned with the choice of forms of propositions for syllogisms. In various ways Aristotle wants to exclude *pas* ("all") from the predicate. As Ackrill 130 says on our *DI* passage, it is unclear whether Aristotle wants to say that a proposition like "Every man is every animal" is false or, as in the *PA*, that it is *achrêston kai adunaton* ("useless and impossible"). His second example in the *PA* is "Justice is every good", without *pasa* ("all") at the beginning—unless that has fallen out of our text. This logical form is also found in his account of the practical syllogism at *EN* 7.3 1147a5–7 where he has as his major premise "Dry food is good for every man", which is then treated as "All dry food is good for every man", followed by two minor premises.

If the scholiast is right in attaching this remark to this passage of Aristotle, what Theophrastus seems to be doing is thinking up an example that falls outside Aristotle's purview. We need not suppose that he took it further and worked out a theory to cover cases like this. But the idea of individual pieces of knowledge is Aristotelian, e.g. *Categories.* 2 1a25–6 τις γραμματική (a piece of grammatical knowledge). See also on **110A**, and on the apparatus to **102A**.

Mignucci 39 suggests that this is a striking advance in logical theory which unfortunately was not taken up by later Peripatetics. It involves taking Phainias and knowledge as the logical subjects of the relation

[65] Aristotle does not use *prosdiorismos* in this sense, but it became the accepted term for "quantifier" in later times. Fortenbaugh 174 n.37 quotes our scholiast as saying that Aristotle called *prosdiorismoi* items like "all" and "no one" (xxxiii 24–5). See n. 58 for details of Zimmermann's table. It could be another example of Theophrastus' interest in naming.

expressed by *echei* ("has"): i.e. seeing having as a two-placed relation. This way of thinking was not of course available in this form to Theophrastus, and can at best have been an intuitive recognition that logic might extend beyond Aristotle's system. It would then be one of the cases in which he puzzled over aspects of Aristotle's teachings.

There has however been a serious challenge to all this: Fortenbaugh argues plausibly that the scholiast may have attached this passage to the wrong place: it should be to *DI* 6 17a33–7. That passage is about contradiction, and if Fortenbaugh is correct Theophrastus is merely adding an apt example, not criticizing Aristotle. The whole thing hangs on whether the scholion is misplaced.

Both Graeser and Repici mangle this passage.

85, 86 The following passages are all concerned with the law of non-contradiction. In all three the commentary is on Aristotle's *Metaphysics* Book Gamma, but in **85A** Alexander refers to Theophrastus' *On Affirmation*, and in **86** just to an *On Affirmation* which, in context, could be a work of Aristotle, but surely must be that of Theophrastus. The implication is that Theophrastus, in his *On Affirmation*, brought in a reference to Aristotle's *Metaphysics*, for only there did Aristotle try to prove the law of non-contradiction.

85A Alexander of Aphrodisias, *On Aristotle's* Metaphysics 4 (Γ).4 1006a11 (*CAG* vol. 1 p. 273.8–19 Hayduck) trans. AlexMet (Madigan) 54–5.

The greater part of this passage is Alexander's account of what he takes Aristotle to be saying in the relevant passage of *Metaphysics* 4 (Γ). Theophrastus is only brought in at the end, and if Alexander stood alone we would know very little.

There is a similarity but not an identity between the words used: Alexander has *biaios* of the proof, while Syrianus says that those who force us to give an argument are *biaiotatous*. It looks as if Alexander is quoting Theophrastus' own words. Whether Syrianus just embroidered those remarks or not is hard to tell. He was in Athens, and had a fine library there. Having read Alexander he might have looked at Theophrastus' *On Affirmation* and drawn on it.

Barnes suggests "from [premises] of this kind" in lines 5–6 and queries "also" for *kai* in line 5, suggesting "actually". As Madigan says, (n. 316) the premises are the respondent's answers.

85B Syrianus, *On Aristotle's* Metaphysics 4 (Γ). 4 1006a11 (*CAG* vol. 6.1 p. 68.26–36 Kroll) (*CAG* has 35 in the margin but it should be 36).

Syrianus died about 437 AD, having been head of the Academy from 432. Most of his work is lost, but this is part of a commentary on *Metaphysics* 2,4,13 and 14.

We have extended this passage beyond what Graeser (Fr.7) and Repici (fr.13b) give, adding our lines 7–11. We cannot be sure that all of this is from Theophrastus, but there is no word in it that he could not have used.

We have put Barnes' suggestion, "stupid" for "unpardonable" in line 4–5.

As we have seen on **85A**, there is a similarity of words between Syrianus and Alexander, whom Syrianus mentions frequently, and Syrianus also seems to be giving us some actual words of Theophrastus. But has he garbled them, or is he just paraphrasing Alexander? The only other place in his surviving commentary where he refers to Theophrastus (upper apparatus to **681** on Archinus' views on the pronunciation of letters), is so close to Alexander (or rather pseudo-Alexander), that it seems obvious that one is quoting from the other.[66] Our passage is much less close. As far as the grammar goes it could be Theophrastus' words to the end of our passage.

The meaning of lines 7–8 could be given in various ways: e.g. "if neither this nor its opposite is more the case". De Lacey, "οὐ μᾶλλον and the antecedents of ancient scepticism", *Phronesis* 3 1958 59–71 argues that *ou mallon* ("no more") and similar expressions were used only non-technically by Aristotle, though by the time of Alexander they had become technical. So if Syrianus is quoting Theophrastus reasonably accurately we need not treat *mêden mallon* ("nothing more") as a technical expression. There is a parallel in Aristotle *Metaphysics* 11 (K) 6 1063b27–8, where οὐδὲν μᾶλλον εἶναι φησὶ γλυκύ ἢ πικρόν is generally

[66] It is still uncertain who the author of the second part of the Commentary on the *Metaphysics* was. Praechter assigned it to Michael of Ephesus of the twelfth century, but Tarán, 'Syrianus and Pseudo-Alexander's Commentary on *Metaphysics* E–N' in Wiesner (1987) 215–50 noting the resemblances to Syrianus, suggested a date more or less contemporary with Syrianus. Bob Sharples, 'Pseudo-Alexander on Aristotle's *Metaphysics* Λ' in *Allesandro di Afrodisia e la "Metafisica" di Aristotele. Temi metafisici e problemi del pensiero antico* ed. G. Movia Milan 2003 199–206, 214 thinks that Pseudo-Alexander used Simplicius and must therefore postdate him.

translated as "he (Anaxagoras) says if nothing is sweet more than it is bitter" or the like.

What *tode* ("this") means is not clear. It could be one part of a contradiction—p as opposed to not-p, or it could be not-(p and not-p) as opposed to (p and not-p). The sense will then be: if it were no more probable that not-(p and not-p) than (p and not-p), then a deductive argument could not prove it, but if it were already more probable then proof would not be needed.

In line 9 we have translated *pragmateia* as "activity". It could mean "subject" as was common later, but that seems too passive here. *karpos* ("fruit") is not used in this metaphorical sense by Aristotle, but it is by Epicurus (*Gnom. Vat.* 27.2 = [6] 27.2 Arr.) and Plutarch, *Adv. Col.* 1117a = [42] 4 Arr.) so that I see no reason why Syrianus should not be basing himself on Theophrastus.

This passage ends with the end of a section in Syrianus, and what follows is not relevant. The question is whether lines 6–11 are from Theophrastus or not. They open with *gar* ("for") and are linked by a further *gar* and a *de* ("and"). So we have an explanation, which may be either Theophrastus' own for his statement that the people asking for an argument are violent etc., or that of Syrianus for his endorsement of Theophrastus' view.

86 Alexander of Aphrodisias, *On Aristotle's* Metaphysics 4 (Γ).7 1011b13 (*CAG* vol. 1 p. 328.14–18 Hayduck) trans. AlexMet (Madigan) 123.

Alexander is commenting on the last section of Aristotle's discussion of the law of non-contradiction. He refers to the *DI* where Aristotle did indeed discuss the matter (Ch. 7–8), but also to *On Affirmation*, a title unknown to us for Aristotle, but for which we have considerable evidence as the title of a work of Theophrastus. Compare the end of **88**. In any case all we get from this passage is that Theophrastus discussed the matter in his *On Affirmation*, presumably in sufficient detail to make it reasonable for Alexander to mention it alongside the *DI*.

87A–F Ackrill. Barnes (1996).

These are all on propositions called "by transposition" by Theophrastus. They were discussed exhaustively by Aristotle, but it was Theophrastus who gave them the name "by transposition" (*kata metathesin* or *ek metatheseôs*). Aristotle does use the related verb *metatithenai* in the neighbourhood of these matters (*DI* 10 20b1, 11), but by him it is used of interchanging the subject and the predicate of a proposition, as with "man is white" and "white is man" (20b1–3). The central problem here is what reason Theophrastus had for introducing the new term; the matter has not seemed to be of great importance to most modern scholars, who have ignored it, but it is of some interest that so many ancient scholars took an interest in it. The central problem is that of distinguishing between not-(A (is) B) and (A (is) not-B). This is easily set out by using a variety of modern notations, but Aristotle had none of these and had to express himself in plain Greek. Theophrastus introduced some new terminology, and in these passages three explanations of it are given.

87A Alexander of Aphrodisias, *On Aristotle's* Prior Analytics 1.46 51b5 (*CAG* vol. 2.1 p. 396.34–397.4 Wallies) trans. AlexPA (Mueller)[4] 93.

This is on the last chapter of *PA* 1, which has been described as a kind of appendix to that book. It relates "the distinctions made in the *DI* about negatives to the theory of the syllogism." In our passage Alexander adds for Theophrastus only that he gave a name to the propositions that Aristotle had already described. The fuller information that follows is from later and perhaps less reliable sources.

The notion of "by transposition" is in fact used rarely in our literature, but there is an example at Alexander *Questions* 2.7 53.10, trans. Sharples,[67] in a discussion of the nature of matter: "it is not the same to say of it 'it is-not, in its own nature, qualified' and 'it is, in its own nature, not-qualified' for 'it is, in its own nature, not-qualified' is an assertion that is said to be 'by transposition' and is equivalent to a privation, but 'it is-not, in its own nature, qualified' is a negation, which does *not* have the same force as a privation, and it is [the *negation*] that is true of matter" (Sharples' translation, with his italics and brackets), Alexander goes on to speak of suitability,[68] which is intermediate between something and its privation. The conclusion is that *not being qualified* is true of matter, but *being without quality* is not. Alexander then quotes parts of *PA* 1.46 51b5ff.

---

[67] Alexander of Aphrodisias, *Quaestiones* 1.1–2.15, London: Duckworth (*ACOA*) 1992.
[68] See Huby, *Commentary on Theophrastus' Psychology* 134 and n. 203.

87B Ammonius, *On Aristotle's* De Interpretatione 10 19b19 (*CAG* vol. 4.5 p. 161.5–11 and 24–32 Busse)

For Ammonius see on **71A**.

The omitted section gives information about Aristotle's procedure. So the subject of the verb in the following sentence (line 7) is Aristotle. Ammonius gives two reasons for Theophrastus' term, either that the order of the propositions is transposed in the diagram, or, quite differently, that the definite predicate is replaced by the indefinite. For the diagram we can refer to *DI* Ch. 10. The fact that Alexander does not give any explanation for the term, but Ammonius and other later writers give alternative suggestions, may imply that the explanation was not known, and that either Theophrastus gave no explanation or that his actual work was no longer available to them. There are indeed certain puzzles here. **87B–E** are all on *DI* 10 19b19, and Ammonius is here talking about *protaseis* or propositions, saying that Aristotle called the one simple and the other indefinite. This opposition cannot be found in Aristotle.[69] He does however at *DI* 3 16b14 speak of indefinite verbs, and at 10 19b10 of indefinite names (nouns), and this might lead later readers to suppose that he spoke of indefinite propositions. See Ackrill 120–1 on indefinite verbs.

What Theophrastus was concerned with was propositions which had negative forms but were in fact affirmations, as Alexander said. Of these it is true that 1) they had a place in the diagram(s) implied by Aristotle, and 2) the definite predicate (man) is replaced by the indefinite one (not-man). Now it is also true that it is immediately after this, at 20b1, that Aristotle himself refers to transposing, but here it is only the transposition of an ordinary subject and predicate. It is difficult to believe that it was this that made Theophrastus use the term as he did, and more likely that it caused some confusion in the minds of later writers.

Ammonius refers to a diagram in Aristotle, which he says he has repeated, but neither in Aristotle nor in Ammonius is there now to be found what we would call a diagram. Ackrill is misleading in translating 19b26–7 "from the following diagram" and then setting out the propo-

[69] At *DI* 5 17a21 he distinguishes between a simple *apophansis* ("statement") and one that is compounded from these, i.e. made up of several simple ones. That is not the distinction we are after.

sitions concerned in a table. All Aristotle says is "from the things written below" (ἐκ τῶν ὑπογεγραμμένων).[70] It is however fairly easy to reconstruct what is wanted, as Ackrill 143 does, and following Ammonius' instructions. So we have:

(A) a man is just	(B) a man is not just
(C) a man is not not-just	(D) a man is not-just

(A) is a simple affirmation, and (B) a simple negation, (C) an indefinite negation and (D) an indefinite affirmation. But it remains unclear what exactly has been transposed. Ammonius' second suggestion, that the definite predicate is replaced by the indefinite is clearer, but is not a very good explanation for Theophrastus' terminology. For more see on **87DE**.

There are two items from John of Italy, for whom see on **77**. He clarifies the issues well: At *Various Questions* 28 (35.26–8 Ioannou) he has "For, when 'not', the negative particle, is added, the predicate term does not make a negative statement, but rather an affirmative, not a simple one, but what those associated with Theophrastus call 'from transposition', and Aristotle 'indefinite'." At 49 (62.4–9) he has "But since 'not' was added here in two ways, with 'to be' as making a negative statement, but with 'just' appearing more like this but not truly so, he (Aristotle) recommends us... to call these an affirmation, but an affirmation from transposition and indefinite, according to Aristotle indefinite, but according to the latter's colleague Theophrastus from transposition."

One might wonder about the text here, but the sense is clear.

There is also an account by Psellus, who was followed by Neophyte, whom Busse showed to be dependent on Psellus. For both see *CAG* vol. 4.5 p. xlix–l. See also K. Ierodiakonou, 'Psellos' Paraphrasis on *De interpretatione*' in Ierodiakonou (2002) 174–7 for other aspects of this matter.

Thomas Aquinas *On Aristotle's* DI.2 *lectio* 2.7 7–13 refers to Theophrastus as the inventor of the name "by transposition".

7C pseudo-Magentinus *On Aristotle's* De Interpretatione 10 19b19 (*CAG* vol. 4.5 p. xliv.11–14 Busse)

Busse thinks this is from the circle of Olympiodorus or his pupils. The genuine Magentinus was of the thirteenth century. See **120**.

At first sight the first and third explanations look the same, but presumably the third means that the position of the proposition in the table is transposed, which is the first explanation given in **87B**. The first here, that the terms of the proposition are transposed, is not clear: is it just

[70] Aristotle uses *diagramma* of geometrical diagrams.

a misunderstanding of what follows in Aristotle? The second, that the word "not" does not remain with (lit. in) its own terms is obscure, but **87D** helps to clarify it.

87D Stephanus of Alexandria, *On Aristotle's* De Interpretatione 10 19b19 (*CAG* vol. 18.3 p. 40.22–8 Hayduck) trans. Charlton (see below) 160.

For Stephanus see W. Charlton's extensive study in 'Philoponus' on Aristotle *On the Soul* 3.9–13 with Stephanus On Aristotle *On interpretation* (translated by Charlton) London Duckworth 2000 (*ACOA)* 1–15. See also Wildberg (1990) note 1, who is doubtful about received views about Stephanus' dates. He thinks he could have lived later than the seventh century, but that should not affect our argument here. See also Blumenthal (1996) 46–7 and 193–4 n. 64.

The first explanation here is that the word "not" is transposed from being with the verb to being with the predicate. This may fit in with the account of Aristotle's remarks about the verb i.e. in *DI* as set out by Barnes 187–92. See Ackrill 117–24 mainly on *DI* 3 for the obscurities in Aristotle's account. At 145 on 10 20b1 he encapsulates the matter: "the meaning of a sentence may be altered if a 'not' is moved from one place in it to another". Theophrastus may have started from this in speaking of transposition, but then brought in Aristotle's account of "indeterminate" terms, perhaps himself applying the word "indeterminate" to propositions.

87E Anonymous, *On Aristotle's* De interpretatione 10 19b19 (cod. Paris. Gr. 2064 = p. 71.1–7 Tarán)

For this work see on **82C**.

Here we have Aristotle contrasted with *hoi peri Theophraston* ("those associated with Theophrastus"). In this case we may accept the standard view that this means no more than Theophrastus.

We have inserted *aoristous* ("indefinite") in line 1 because it seems required by the sense and its loss can easily be accounted for.

The last word, *hepomenai* ("follow"), is not in Ammonius It may be from another source, possibly Syrianus, which Tarán recognises as being here. Or, as we suggested on **82C**, there might have been a handbook. Hintikka (1973) 53–5 distinguishes various senses of *hepesthai*. Here "be compatible with" is probably correct.

87F Ibn-Buṭlān in: Ibn-Abī-Uṣaybiʿa, *Essential Information on the Generation of Physicians* 14, chapter on Ibn-Riḍwān (vol. 2, p. 102.18–20 Müller)

J. Schacht and M. Meyerhof, The Medico-Philosophical Controversy between Ibn Buṭlān of Baghdad and Ibn Riḍwān of Cairo, Cairo 1937, p. 53 (Arabic text), 85 (English translation); Gutas (1985) 90–91.

This passage is taken from the correspondence of Ibn-Butlān addressed to ʿAlī Ibn-Ridwān, a correspondence that records the controversy that took place between the two physicians in 1050 AD. ʿAlī Ibn-Ridwān, who was a Muslim Egyptian of low birth and grew up in Cairo under impoverished circumstances, learned medicine on his own from studying books. His adversary, Muḫtār ibn-Buṭlān—the author of this passage—was a well to do Christian of Baghdad where he studied medicine with several scholars, most notably the philosopher and physician Abū-l-Farağ ibn-aṭ-Ṭayyib, the author of the commentaries cited at **71D** and **71H**. In this passage, Ibn-Buṭlān makes the argument that correct education can only be had from teachers, not from books alone, for it is the teacher that will correct any mistakes in the handwritten books and interpret their contents—an argument intended to denigrate the knowledge of the autodidact Ibn-Riḍwān. In the course of his argument, Ibn-Butlān refers to Theophrastus and Eudemus as an example of the point he is making. For the nature and causes of the controversy see Schacht & Meyerhof, whose translation of the passage I follow with slight modifications. The text I translate is taken from Ibn-Abī-Uṣaybiʿa, the historian of physicians (d. 1258), who quotes from this correspondence in what is apparently a slightly superior textual tradition.

Ibn-Butlān, who studied in Baghdad with the Baghdad Peripatetics, had access to a wealth of material on the Greek Aristotelian commentatorial tradition in Arabic translation. It is clear that he draws his information from such a source. That no mention of the particular point he makes is extant in any of our Greek sources would indicate, first, that this source was in all likelihood not a generic commentary on *De Interpretatione* but a specialized discussion of the difficult passage 19b19ff., and second, that it was of a relatively late date (sixth century?) because it did not find its way into the general commentaries. (DG)

This passage is different from any of the others both in the point that it is making and in that it is the only one to mention Eudemus. (PMH).

88 Alexander of Aphrodisias, *On Aristotle's* Prior Analytics 1.37 49a6 (*CAG* vol. 2.1 p. 367.7–14 Wallies) trans. Alex PA (Mueller)[4] 57.

At *DI* 11 Aristotle discusses cases like "white walking man", "good cobbler", and "white musical" which involve the question of the unity of combined predicates, but is far from lucid.[71] Alexander here, on the very short chapter 37 of *PA* 1, refers to that discussion but adds that Theophrastus said more in his *On Affirmation.* That may be one of the places, referred to by Boethius at **72A**, where Theophrastus elaborated on Aristotle, but we do not know what he said. At **134** Alexander quotes *On Many Senses* for something similar.

89 Alexander of Aphrodisias, On Aristotle's *Topics* 3.6 120a20 (*CAG* vol. 2.2 p. 290.1–2 and 9–10 and 24–9 Wallies)

Brunschwig (1982).

This is not very informative. The last words even appear to show that Alexander was not familiar with the text of Theophrastus' *On Affirmation* at this point. Perhaps his copy was mutilated at the end. In any case he does not tell us what Theophrastus said.

Brunschwig notes that this is an advance on the passage of the *Topics* (3.6 120a27–31) at which Alexander quotes it. There Aristotle lists four ways of refuting a certain thesis: here there are five ways. It remains unclear, however, whether Alexander is giving Theophrastus' account.

At the beginning Alexander appears to distinguish between "problem" and "thesis" as if Aristotle's way of speaking were unfamiliar. Aristotle himself covers the question of the relationship between the two in the first book of his *Topics* 11 104b18–105a2, saying that all theses are problems, but not all problems are theses. No doubt Theophrastus followed him. See on **135**.

At **118** Boethius discusses *DI* 5 17a8 and refers to Theophrastus *On Affirmation* about definition.

It has been suggested by Busse (*CAG* 4.5 p. vi n. 2) that chapter 14 of Aristotle's *DI* is not by Aristotle but could be the work of another Peripatetic, possibly Theophrastus. This chapter was ignored by Porphyry

---

[71] At *Metaph.* 8 (H).6 1045a7–14 Aristotle assimilates the unity of numbers to that of definition (*horismos*). But he then goes on to the unity of e.g. man.

and perhaps not known to Andronicus. See Prantl **I** 359. Its doctrine and its position at the end of the work also suggest that there is something unusual about it, but Ackrill 153 accepts its Aristotelian origins, though he thinks it does not belong to the *DI*. At 70 he suggests that it was originally an independent essay or lecture. See now also De Rijk (2002) 347–8.

Categorical Syllogism

Much of our evidence comes from AlexPA, and is from Theophrastus' own *Prior Analytics*. It is unlikely that that work was written until Aristotle had left Athens in 323, and most likely not till after his death in 322. That does not mean that the ideas it contains were not formulated till then. Theophrastus must have shared in Aristotle's thinking over many years; but we may think of him as engaged with Eudemus and others in organising Aristotle's "papers" and perhaps at that time formulating some of his own ideas. But it is also likely that, to some extent with Eudemus, he had worked on modal logic earlier and, as I shall suggest, provoked Aristotle into some replies.[72] Eudemus is said to have competed with Theophrastus for the headship of the school (See **8**), and probably soon afterwards left Athens for Rhodes where he set up his own school. He did correspond with Theophrastus, and there is no reason to suppose that he could not have returned at intervals to Athens, but he is unlikely to have worked closely with Theophrastus in later years.

It is natural that what Alexander and others have recorded for us is about where Theophrastus differed from Aristotle. We can only therefore guess what form his complete works took. If Boethius (**80**) is a guide he may have given a basic account of syllogistic theory, and relied on his readers to consult Aristotle's own work for details.[73]

There is however one passage that suggests something different. That is **94**, from Averroes, which covers a complicated approach to the appropriateness of the ways in which premises are formed according to the figures they are in.

---

[72] Bocheński 125 argues that in much of his work Theophrastus is merely writing up material already worked out by Aristotle. But it seems that every scrap of writing left by Aristotle was painstakingly added to his corpus by his "editors". Repici 41–2 surveys views that have been held about the connection between Theophrastus and Eudemus.

[73] **121** gives an Arabic version of an entry in a list of Aristotle's works. It suggests that Theophrastus dealt with definition in his *Prior Analytics*.

90AB These are on the conversion of simple universal negative propositions like "No B is A", to "No A is B". Conversion played an important part in Aristotle's syllogistic, and at *PA* 1.2 25a14 he gives a proof of the above conversion by means of a form of *ecthesis*,[74] in which an item C is considered as part of A; *ecthesis* has been a stumbling-block to many logicians, as they (Theophrastus and Eudemus) seem to have agreed. So they produced a proof by means of the idea of separation, i.e. that if no As are B, then all As are separate from all Bs. At AlexPA (Barnes) 86 n. 14 it is pointed out that the idea of separation was adopted by Boethus and then Maximus in connexion with the argument that all syllogisms are perfect. See **94**.

90A Alexander of Aphrodisias, *On Aristotle's* Prior Analytics 1.2 25a14 (*CAG* vol. 2.1 pp. 31.4–10 and 34.13–15 Wallies) trans. AlexPA (Barnes) 86 and 89–90.

Flannery 27–8 n. 69 (for lemma). Mignucci (1998) 46–9.

In the gap between our passages Alexander examines Aristotle's proof and various interpretations and criticisms of it.[75] But lines 34.8–22 are, as is said in AlexPA (Barnes) 89 n. 29, repetitive and inconsistent, and "probably represent an earlier version of Alexander's commentary".

We have kept the singular form *onomaze*, which has better manuscript support, and we take[76] it that its subject is Aristotle, and that in the translation we should substitute for "he" the words "for here he (Aristotle)", and remove the note; the expression "*universal privative*" is indeed at *PA* 1.2 25a14 and the sentence with *gar* (for) in lines 2–3 is parenthetical.

We have used "disjoined" for the Greek "*apezeuktai*", which is the word most frequently used in this context by Alexander. He does once have "*kechôristai*", for which we have used "separated": that is the word adopted by Philoponus in **90B**, but it seems likely that Theophrastus and Eudemus used the more unusual word "*apezeuktai*" ("disjoined").[77]

[74] See K. Ierodiakonou, 'Aristotle's use of examples in the *Prior Analytics*' *Phronesis* 47 (2002) 150–1.

[75] See AlexPA (Barnes) 87 n.19 for a suggestion about Theophrastus' criticism of Aristotle's way of dealing with the conversion of universal negative propositions.

[76] Aristotle does not use *chôrizein* and related words in a logical context. Alexander uses it at 26.19–22 in an elementary account of Aristotle's original distinctions.

[77] Patzig 163 says that Alexander rebuked Theophrastus and Eudemus for saying that Aristotle's proof was circular. But Alexander (31.27) speaks only of *tines* ("some") in

Mignucci 47–9 and nn. 32–3 considers the objection raised by some that Theophrastus' proof is not logical but spatial, and suggests a way of interpreting it without modern technical complications.

Mignucci 44 n. 15 has a full bibliography on ecthesis.

Additional text

The following is to be added to the upper apparatus of the text:

A parallel text by Alexander to lines 1–6 of this fragment is found in his essay, *On the Conversion of Premises*, in Badawī (1971) p. 65.1–5 (for this essay see below at **91C**). The text there reads:

> Aristotle's colleagues, Theophrastus and Eudemus, gave a simpler account of this conversion [i.e., of the universal negative] than this. The reason is that they said that, if A is predicated of no B, then it is disjoined from it and separated from it; but what is separated is separated from that from which it is separated; so B [too] is separated from A; and since [B] is separated from [A], then [B] is <predicated>[78] of no [A]. (DG)

B Philoponus, *On Aristotle's* Prior Analytics 1.2 25a16 (*CAG* vol. 13.2 p. 48.11–18 Wallies)

Mignucci (1998).

For Philoponus see on **71E**.

Philoponus uses *chôrizein*, "to separate" throughout. He also enlarges the argument by saying that A is separated from all the parts of B, and brings in the notion of "*pros ti*" or relative. In line 7, in the second impression, we have substituted "what is separated" for "'separated'", but in any case it is not easy to distinguish between words and things here. I am inclined to think that this point was not one of Theophrastus' original arguments.

Here we may refer to AlexPA 100.17–26, where Alexander is considering *PA* 1.6 28a15–26, which includes three proofs of Darapti. He recounts Aristotle's proof by *ecthesis*, and says that what is "exposed"

---

this context, after a reference to Theophrastus and Eudemus by name, and is unlikely to be referring to them in this way here.

[78] Adding here <*tuḥmalu*> to read, *fa-laysa hiya* <*tuḥmalu*> *ʿalā šayʾin minhā*, in order to parallel the earlier construction (line 3), *lā tuḥmalu ʿalā šayʾin min B*.

is one item, which is perceptible. He goes on to refer to "some" (*tines*) who think that with one item they will find the predicates (of the syllogism and of the exposed item) either linked (*sunêmmena*) or separated (*kechôrismena*) from one another. Flannery (37–8) argues that these people may be Theophrastus and Eudemus (and their followers) pointing to the respectful tone used and to the similarity with the argument of **90**.[79]

91A–D Bocheński. Lukasiewicz 23–30. Rose. Patzig 109–18. Mignucci (1998). AlexPA (Barnes)

Aristotle set out his theory of the categorical[80] syllogism in his *PA*. He had three figures, according to the position of the three terms in a syllogism, and had the idea of a perfect syllogism, i.e. one that was obviously true and needed no proof. For him the first figure contained only four syllogisms, all perfect. These are the standard ones

All Bs are A
All Cs are B
Therefore All Cs are A. (Barbara)

No Bs are A
All Cs are B
Therefore No Cs are A. (Celarent)

All Bs are A
Some Cs are B
Therefore Some Cs are A. (Darii)

No Bs are A
Some Cs are B
Therefore Some Cs are not A.[81] (Ferio)

[79] He refers however also to the possibility that Boethus may be meant, citing AlexPA (Barnes) 174 n. 37.

[80] The term "categorical" is not used in this way by either Aristotle or Theophrastus. Galen 7.9 (19.8–18 Kalbfleisch) takes the view that "categorical" has by custom been used of both affirmative and negative premisses, and then applied to syllogisms containing them. See Kieffer 103. Kieffer points out that the expression "categorical syllogism" was only needed after the development of the theory of hypothetical syllogisms. See now Bobzien (2002a) 361–4 and Mueller (2006b) 6.

[81] For more on Aristotle's syllogistic see pp. 3–4.

By the use of conversion Theophrastus and Eudemus[82] produced five more syllogisms which are not perfect. These are mentioned by Aristotle himself, but not organised in a figure (*PA* 1.7 29a19–27 and 2.1 53a8–12).[83] Mignucci assumes that Theophrastus just went through Aristotle's work and assembled the cases mentioned here and there. But this supposes that what Theophrastus was working on was the *PA* as written and left behind by Aristotle. But if he was working with Eudemus, it is likely to have been in Aristotle's lifetime, and there is no reason why they should not have taken the fairly simple but systematic step of having a clear arrangement. We know from **90** that they were interested in the conversion of premises: why not take it further? A different view comes from Lukasiewicz 27, following Bocheński 59. They think that 29a19–27 and 53a8–12 are in passages added later by Aristotle who did not get around to systematizing the remaining cases, but left the job to Theophrastus. Lukasiewicz suggests that Theophrastus changed the account of the first figure from one in which the middle term is the subject of the major and the predicate of the minor premise, (1.23 41a13), to one in which the middle term is just the subject of one premise and the predicate of the other. That is the position stated by Alexander (on 1.23 at 258.17 and on 1.32 at 349.5), without mention of Theophrastus, and without, according to Lukasiewicz, noticing that it differs from Aristotle's. Mignucci (43) also criticizes Theophrastus for omitting the "subaltern" moods, which have a particular instead of a universal conclusion. But that is to assume that Theophrastus' purpose was to hoover up Aristotle's sidelines. See also on **92**.

It is clear from what Alexander says that Theophrastus arranged the syllogism in a carefully thought out order, whereas Aristotle seems uninterested in that matter. It is, then, possible that *PA* 1.7, which, as we have seen, Bocheński 59, Lukasiewicz 27, and others have thought a later addition by Aristotle was indeed a later addition under the influence of discussions with his colleagues.[84]

---

[82] Alexander mentions only Theophrastus; Boethius (**91D**) seems uncertain with *vel* (or) in line 2 and only Theophrastus in **91E**. Apuleius (**92**) has "Theophrastus and the rest" and Aratus (**93**) names both.

[83] We do not have Alexander's commentary on the second book of *PA*, so that we do not know what he said on this passage. But his straightforward comments on 29a19 (**91B**) take us no further.

[84] Patzig 109–13, discusses Aristotle and the fourth figure, and lists what he takes to be Theophrastus' departures from Aristotle's system.

In **91A–C** Alexander attributes these five extra moods to Theophrastus only, but in **91D** Boethius has "Theophrastus or (*vel*) Eudemus" and "Theophrastus and Eudemus".

At AlexPA (Barnes) 136 n. 157 it is said that no text hints that Theophrastus added anything to the other figures.[85]

91A Alexander of Aphrodisias, *On Aristotle's* Prior Analytics 1.4 26b30 (*CAG* vol. 2.1 pp. 69.26–70.21 Wallies) trans. AlexPA (Barnes) 135–6

In the second impression in line 2 we have substituted "are the primary members of" for "took the first places in", as suggested by Barnes.

In n. 2 put Celantes.

91A and **91B** should be taken together. They make it clear that whereas there was a later development that placed the five syllogisms in an extra, the fourth, figure, Theophrastus' innovation was to add them to the first figure in a definite order. It is clear that both Aristotle and Theophrastus saw these five moods as extensions of the first figure.[86]

Here Alexander deals with a) Barbara, where by conversion of the conclusion All Cs are A we get Some As are C (Baralipton); b) Celarent, converting No Cs are A to No As are C (Celantes); c) Darii, with Some Cs are A being converted to Some As are C (Dabitis). The remaining two have more difficult operations. They involve the conversion of both premises: thus the eighth has:

All Bs are A 1)
No Cs are B 2)

Convert 2)	No Bs are C
Convert 1)	Some As are B

And we get:

Therefore Some As are not C (Fapesmo)

[85] Th. Ebert, "Warum fehlt bei Aristoteles die 4 Figur?" *Archiv für Geschichte der Philosophie* 62 (1980) 15 thinks Aristotle did not have a single view about the figures, and differed between *PA* 1.4–6 and 1.7 (and 2.1). Theophrastus therefore modified Aristotle's system leaving open the possibility of a fourth figure.

[86] AlexPA (Barnes) 213–5 has a table of additional syllogisms.

And the ninth has:

Some Bs are A 1)
No Cs are B 2)

Convert 1) Some As are B
Convert 2) No Bs are C

And we get:

Therefore Some As are not C (Frisesomorum).

Mignucci (1998) 42–3 discussses this material but concludes that here at least Theophrastus' work is of little originality.

Schol. On Aristotle's PA. 2.1 53a4 in various codices = p. 188a4–12 Brandis

This is Repici fr. 23g. The scholiast says that Theophrastus had nine syllogisms in the first figure, and explains where they are to be found in Aristotle's work.

B Alexander of Aphrodisias *On Aristotle's* Prior Analytics 1.7 29a19 (*CAG* vol. 2.1 p. 110.12–21 Wallies) trans. AlexPA (Barnes) 185–6.

This repeats **91A**.9–17.

In lines 3–5 put "do not at all prove what was propounded by conversion of the conclusion as do the three before them" (Barnes).

Alexander's remarks on *PA* 2 1 53a10–14 are lost with his second book.

C Alexander of Aphrodisias, *On Aristotle's Conversion of Premises* (p. 60.8–9 Badawī).

DPA I 136, no. 28. Josef van Ess, 'Über einige neue Fragmente des Alexander von Aphrodisias und des Proklos in arabischer Übersetzung,' *Der Islam* 42 (1966) 149–54. F.W. Zimmermann and H. Vivian B. Brown, 'Neue arabische Übersetzungstexte aus dem Bereich der spätantiken griechischen Philosophie', *Der Islam* 50 (1973) 313–324. Bocheński (1947) 56–58; Rosenberg & Manekin (1988) 97–98

For Alexander's essay, extant only in Arabic, see van Ess, no. 36, Zimmermann & Brown, 320–321, and the entry in *DPA*. The edition of the Arabic text by Badawī (1971) 55–80, based on the so far unique manuscript in the Public Library in Tashkent (no. 2385), was made from enlarged photographs of the microfilm kept in the Bibliothèque Nationale in Paris (see G. Vajda in *Arabica* 21 [1974] 219); it is reviewed by R.M. Frank in *Journal of the American Oriental Society* (1974) 245–7. The passage in question referring to Theophrastus is discussed by Bocheński. (DG)

This is a terse and not entirely accurate summary of the position of the five extra syllogisms. (PMH)

91D Boethius, *On Aristotle's Categorical Syllogism* 2 (*PL* vol. 64 col. 813B–C and 814C and 815A–816C Migne).

colligere is here rendered "draw" (27,61), "imply" (17,37), "deduce" (35,39,44,46,64,71), and "derive" (50,77) because of the peculiarities of the English language. In line 64 *praedictur* has been corrected to read *praedicetur* in the second impression.

Here Boethius goes through all five of the extra moods in a straightforward way. We do not know the source of this material. He twice (lines 65 and 81) names the *Resolutoriis*, which is Latin for *Analytics*. The explanations he refers to are in Aristotle's *PA* 1.4 26a3 and 37. He could be thinking of his own translation of the *PA*, now lost, or of a commentary on it, also lost. It has also been suggested that he was just translating a Greek work. But he does say *diximus* ("we have said") in the second case. Note that in lines 4,8, and 11 he refers to Aristotle's own work as *Analyticorum* etc. See on **91E**.

Line 14 Boethius uses the Greek expression κατὰ ἀνάκλασιν followed by its Latin equivalent *per refractionem* to describe these added syllogisms, and in AlexPA (Barnes) 136 n. 157 it is suggested that the Greek term may go back to Theophrastus. See on **95A** where the related term *reflexim* is found.

Abaelard, *Dialectica* 2.3 p. 238 has taken the account of the five moods from Boethius.

Ars Burana 3 (199.20–2 and 201.19–20) just says that Eudemus and Theophrastus added the five moods.

91E Boethius, *On the Categorical Syllogism* 2 (*PL* vol. 64 col. 829D Migne).

Boethius must have used Porphyry directly, and it could be that he got his information about Theophrastus indirectly through Porphyry.

Barnes has suggested that **93** might appropriately be treated as **91F**.

92 Apuleius, *De Interpretatione* 13 (*BT* p. 193.7–13 Thomas) trans. Londey and Johanson 104.

Sullivan 1967. Huby, 'Apuleius and Theophrastus' fifth "indemonstrable" mood', *LCM* (1977) 147–8. D. Londey and C. Johanson, *The Logic of Apuleius* Leiden: Brill 1987. Barnes 2002 106.

This may be our earliest evidence for the logic of Theophrastus, and also the first work on formal logic to have survived in Latin. Its authenticity has been doubted, and scholars are still divided about it. If it is genuine, it is likely to have been written about 155 when Apuleius was attempting to pass on in Latin the fruit of his studies in Greek. It may be seen as a continuation of his *On Plato and his Teachings.* There is however a difficulty about its use of the word *modus*,[87] which also occurs in **95A**. As is shown on **106F**, the Greek equivalent, *tropos*, is not found with the sense of "mood" until it appears in pseudo-Ammonius in the sixth century. Earlier writers handle the subject perfectly well without it. It is tempting to suppose that the work is not by the second century Apuleius, but much later. *Modus* in the sense of "mood" was however used by Martianus Capella, probably of the fifth century, which is earlier than pseudo-Ammonius.[88]

An alternative view is that the systematized logic found in Apuleius was developed earlier than pseudo-Ammonius, as part of an educational development ignored by the superior thinkers of the Commentaries on Aristotle. Handbooks were undoubtedly made to introduce students to elementary logic. In them the word *tropos* would have found its own peculiar sense, that with which we are now familiar. Apuleius could well have come across it when in Athens in 155. Naturally there is little evidence for such works. One that has survived by luck is Galen's *Introduction to Logic*, written probably late in his career and therefore about 200 AD, which is too late to be a source for the genuine Apuleius. In any case Galen follows Aristotle in speaking of syllogisms, rather than moods, in the figures, but at 6.6 he says that logicians use the word *tropos* for the schemata of arguments, applying it however to Stoic argument forms only. Kieffer 20 discusses the evidence from Aulus Gellius and Proclus for an industry among the Stoics for the production of handbooks.

---

[87] See Londey and Johanson 58–9 for remarks about moods in Apuleius.
[88] At 4 327.7 (151.5 Dick) he has: *modos demum bis quinque profatus* of Aristotle.

It would however be only a small step to apply the word *tropos* to the schemata of the Peripatetic syllogisms.

It has been suggested that, whether it is authentic or not, Apuleius' work is a translation of a Greek original. Scholars have been impressed by the attempts to find Latin equivalents for Greek terms. See on **95A**. In any case there is no reason to doubt the reliability of the evidence it contains.

This passage has been vindicated against Bocheński 16 n. 22 by Sullivan 156, who shows that Apuleius is using the sense of "indemonstrable" mentioned at **188.5–11** which means "so obvious as to need no proof". Unfortunately there seems to be a lacuna here which covers Apuleius' further explanations, but Sullivan has shown that the mood concerned must be Darii or Ferio, and Huby has argued that only Darii is suitable. For Darii we have, using Apuleius' way of formulating the syllogisms:

Everything B is C
Some A is B
Therefore Some A is C.

We then get, substituting "A is B" for "Some A is B":

Everything B is C
A is B (indefinite)
Therefore A is C (indefinite).

That case could be seen as obvious, whereas a similar treatment of Ferio would not. For we would have:

No B is C
Some A is B
Therefore Some A is not C.

and then by the same substitution

No B is C
A is B
Therefore A is not C.

Not only does this not seem obvious, but at **82A** Alexander says that Aristotle treated "not belonging to some" as indefinite and implies that

Theophrastus took the same line. See also **83**. So "Some A is not C" would itself already be indefinite.

Even with the text as it is it is clear that Apuleius did not think much of this innovation, and it is puzzling why he bothered to mention it in this comparatively short work.

Philoponus, *On Aristotle's* PA 95.10–17 and 111.4–23, lists moods in the second and third figures with, in the second, the two final moods out of six having one indefinite (*aprosdioristos*) premise, and, in the third, the four final ones out of ten also having one indefinite premise. Note that he does not speak of *tropoi* but of syllogistic and asyllogistic combinations. Barnes 106–8 appears ro believe that Theophrastus may have said more than our inadequate evidence allows, and considers some later developments in this field.

We have used "the rest" for *ceteri*, but "others" would also be possible. Perhaps Eudemus was among them.

93 Aratus, *Philosophy*, cod. Paris. Gr. suppl. 645 f. 197r (*Jb. class. Phil.* suppl. 23 [1897] p. 707.5–9 Kalbfleisch) trans. H.B. Gottschalk in *PCSB* vol. 3 258 (for source see ibid. xiii).

This work is also to be found in another MS, Moscow 1894 VI 457 p. 694, containing Aratus' *Philosophy*, an account of Aristotle's complete logic. It refers again to Theophrastus in a forward-looking reference to his being the father of the fourth figure. This MS has been studied by the late Hans Gottschalk, for whose help I am very grateful. He wants to put Aratus fairly late. The latest names mentioned are Ammonius and Simplicius. The name of the author, Aratus, is only to be found in the Moscow MS. The work gets more sketchy towards the end. It shows knowledge of Western developments in the theory of the moods of the syllogism, which are to some extent worked out on medieval lines. The MS is of the eighteenth century, and Barnes has suggested that the material could be of the sixteenth to seventeenth centuries. See also **102C**.

These lines are part of an *aporia* (problem): "besides these it seemed to Aristotle and those around him[89] that there could not be another syllogistic figure; for it was impossible in any other relation apart from those

[89] τοῖς περὶ αὐτόν: see n. 47. This could mean just Aristotle, and his closest associates, Theophrastus and Eudemus, are certainly excluded, because they are in opposition to him. In the *apantêsis* ("reply") there is mention of the genuine *thiasôtai* ("companions") of Aristotle, who always agree with him. We do not know who these were.

mentioned for the middle term in three terms to be arranged in relation to the two extreme terms;" then comes our passage and after it:

> For the establishment of this they use the following argument: if it is possible for the middle term to be arranged with regard to the extremes in another relation, there will be another syllogistic figure besides those mentioned; but the first; the second also then. The minor has been demonstrated, for it is possible for the middle term to be predicated of the major and to be the subject of the minor, contrariwise to the first figure, and let it be subject to the minor, like:
>
> Every man is animal,
> Every animal is substance
> Then, Some substance is man
>
> Man is the major term, because in the conclusion it is the predicate, and substance is the minor, because it is the subject in the same. Animal, which is the middle term, is predicated of man, and as the minor extreme it is the subject of substance.

A reply (*apantêsis*) to this is given, in effect that Aristotle's figures are correct, and the new case is just one of interchanging the order of the premises.

Now clearly much of this is late, and uses Stoic argument forms.

The younger men who attributed the fourth figure to Galen must postdate Galen, but that is all we know. See on **110C**. Galen himself knows nothing of a fourth figure, and indeed at *Introduction to Logic* 12.1 he denies that more than three figures are possible; at 11.3–7 he does discuss related material, but with no mention of Theophrastus. See after **95B**. Lukasiewicz 38–42 discusses the fourth figure at length, and suggests that what Galen did was to have four figures of compound syllogisms, i.e. those with four terms, and in a muddled way this led to the attribution to him of the fourth figure of the simple syllogisms. He quotes from a scholium printed by Wallies in the introduction to his edition of Ammonius' *On Aristotle's* PA 1 p. ix. It refers to Galen's *Apodeictic* for this account. That may be the same as Galen's memoranda on demonstration of our **114**, which he also refers to in his *Introduction to Logic* 17.1 as being an earlier work.[90]

Barnes has suggested that this might be treated as **91F**.

[90] For a wise discussion of the fourth figure see the Kneales 101. Hans Gottschalk (1987) 1171 deals adequately with the claim by Rescher to have found new evidence in Arabic sources.

Additional reference

At Cod. Par. Suppl. Gr. 645 f.200r we have:

> τρεῖς ἑτέρας ἐκτίθησι ὁ Θεόφραστος ἐπὶ τοῦ πρώτου σχήματος. διαφέρουσι δὲ οὐδὲν τῆς πρώτης καὶ τρίτης καὶ πέμπτης πλὴν ὅσον ἀντεστραμμένως συνάγουσι τὸ ἔλαττον ἄκρον κατὰ τοῦ μείζονος.
>
> Theophrastus sets out three other (combinations) in the first figure; they differ from the first and third and fifth in no way except insofar as they conclude the minor term of the major conversely.

The writer gives examples in two forms, beginning either:

> Ensouled of every living thing, or,
> Every living thing (is) ensouled.

According to Wartelle (1963) 119 this is an eighteenth century MS containing logical commentaries on Aristotle. Since it is so late, it is unlikely to give us any extra information about Theophrastus. The passage contains problems of interpretation, such as what the fifth (combination) was, but these are beyond the scope of this work.

94 Themistius, *Against Maximus, On the Reduction of the Second and Third Figure to the First* (p. 314.5–6, 324.16–325.22 Badawī 1947).

Themistius' treatise, extant only in Arabic, is translated into French by Badawī (1968), pp. 170, 179–80.

Correction: in the apparatus criticus of the Arabic text of this passage, the line numbers are off by 2; thus line number 17 should be 19, 21 should be 23, etc. (DG)

T-S. Lee (1984) 128–9. G. Striker (1996) 214–9 discusses Maximus.

For Themistius' unusual philosophical position see Blumenthal (1996) 22–3, 37–8.

This passage is good evidence that Theophrastus paid attention to the peculiarities of some of the premises involved in certain syllogisms according to Aristotle's rules. And it seems likely that the whole of the argument given here derives from him and Eudemus. The classification into appropriate, accidental, and unnatural types of predication, and the discussion of their relevance to the figures of the syllogism all hang together.

Where did Themistius obtain this information? Since both Theophrastus and Eudemus are named it is not likely to be from either of Theophrastus' works named in **97**, but more probably from something connected with the *PA*. On the other hand it shows some concern with students, but Aristotle does that too, as in *PA* 1.32 onwards. Very few syllogisms cannot be reduced to the first figure, so that one may ask what is the point of the other figures. One answer is that some propositions are more natureal in the other figures: this clearly takes account of the needs of students rather than of high-powered logicians.

The explanation of accidental predication is clear from the example, "The white thing walks", and this example is related to a number of places in Aristotle. There are however problems: some of Aristotle's cases are in the form of singular propositions, like "That white thing is Socrates' at *PA* 1.27 43a35 which in certain situations is a plausible example of accidental predication but cannot be part of a canonical syllogism. The white swans example is one of Aristotle's standbys.[91] It is unfortunate that we have this section only in Arabic, so that we cannot know exactly what form of Greek was used.

The next example seems to amount to this:

> All stars move in a circle,
> No water moves in a circle,

with the implied conclusion that

> No stars are water.[92]

The objection to a first figure syllogism was that it would have as a premise

> Nothing that moves in a circle is water,

which would seem unnatural. The second figure syllogism would therefore be better at least from the point of view of less able students.

The example in the third figure appears to be that either premise would be unnatural if converted, i.e. both

> Some white thing is a swan, and
> Some musician is a swan.

---

[91] *PA* 1.16 36b11, 19 =, b20, 2.2 54a21.

[92] Earlier fire has been mentioned instead of water, but for the purposes of the argument the difference is not significant.

We cannot be sure how much of this passage is evidence for the views of Theophrastus and Eudemus. Only the part about unnatural predication is certainly so, but it fits well into the whole argument for the value of the second and third figures.

Themistius says that Theophrastus was not concerned with the generation of the second and third figures from the first, so that we need not pursue this matter.[93] (PMH)

Ammonius *On Aristotle's* PA 1.1 31.11–25. The text is partly in Repici fr.25, and there is a translation by Sorabji in *PCSB* 3 259.

Patzig 69–78

This is a concise account of views held on either side of the dispute about perfect syllogisms. It starts with Boethus, described as the eleventh[94] from Aristotle, who is said to have proved that all the syllogisms in the second and third figures are perfect. He was followed by Porphyry and Iamblichus, and also Maximus. Themistius was of the opposite opinion, and there follows an account of the argument between him and Maximus, and of Julian's decision. It is at this point that there is a sentence about Theophrastus, after which Ammonius goes back to Proclus and his own father Hermeias, who are said to have followed "those from (*apo*) Boethus'.

Patzig explores the notion of perfection at length, but does not mention Averroes. It appears too that he did not know the Arabic version of Themistius which we have here, and so only deduced Theophrastus' views from what he knew of Ammonius. The sentence about Theophrastus is not only in an odd position: it is strange in itself. It says that it is clear that he held (*phainetai . . . echôn*) the opposite view on this matter to that of Aristotle. But no evidence is given about this. Lee 128 n. 18 remarks that Ammonius seems uncertain, and that Theophrastus' work had probably disappeared by his time. Bocheński 22 n. 66 shares this opinion. We cannot pursue this matter further, but it is puzzling.

We must however bring in **Brandis cod. Par. Schol.** 156b44 which reports of Maximus that he argued: Consider, M to none of the Ns; M to all the Xs; therefore N to none of the Xs. Since the M is of all the Xs

---

[93] For more on this see Henry Maconi, 'Late Greek Syllogistic' in *Phronesis* 30 (1985) 96–8.

[94] He succeeded Andronicus as scholarch of the Peripatos. The numbering is uncertain.

but none of the Ns the N is separated then from the M: will not the N be separated entirely from the X and belong to none of it? And see, we did not need the reduction to the first figure." This argument is on the same lines as that used by Theophrastus and Eudemus in the **90**s.

Moraux, (1984) 168, referring to Bocheński 64–5 has suggested that Aristotle himself came round to the view that all syllogisms are perfect. It is true that in a number of places (*PA* 1.44 50b9, 2.8) he does discuss reducing the first figure moods to those in other figures, but that seems to be no more than part of a number of logical "exercises" in which he engages. Lynn Rose, on the other hand (1968 127) suggests that Theophrastus distinguished between "direct" and "indirect" moods in all figures, and one might then be able to claim that the direct moods were in a certain sense perfect in all figures. Patzig 74–5 notes that Alexander (77.6–9 and 26–8) says that it is reasonable to call syllogisms which need two conversions more imperfect than those which need only one, and suggests that that derives from Theophrastus as in **96**.

95–6 Both of these are concerned with the third figure, where Theophrastus is said to have made a number of changes, adding one syllogism and changing the order of others. Since he is elsewhere said to have made changes for didactic purposes, we must consider that possibility here. We must also take note of the information derived from Themistius given in **94** and that given by Alexander in **97**.

Aristotle's order in the third figure is:

Darapti, Felapton, Disamis, Datisi, Bocardo, Ferison—six in all. Theophrastus is said to have added Daraptis, for more on which see **106F**, and to have interchanged Datisi with Disamis, and Ferison with Bocardo.

95A Apuleius, *De Interpretatione* 11 (*BT* p. 189.19–27 Thomas) trans. Londey and Johanson 98–101.

Sullivan 1967

For Apuleius see on **92**. In his exploration of the third figure Theophrastus is said to have argued for there being two separate syllogisms in the first mood (Darapti): either the major or the minor premise may be converted and used with the remaining one to give a syllogism in Darii with a particular conclusion. Alternatively the original conclusion may be converted to give the same result. The idea that there are two syllogisms in Darapti has been severely criticized and discussed at length; see AlexPA (Barnes) 168 n. 18. It is therefore worthwhile to speculate on what Theophrastus might have been doing. In **95B** and **96** it is implied that he followed Aristotle in having six moods in the third figure, although he changed their order.[95] On this evidence he did not slip in a second Darapti between the first one and Felapton. Further, in **95B** it is only "some people", not Theophrastus by name, who are said to have made the number here seven. It could therefore be that Theophrastus discussed Darapti elsewhere than as part of the list.[96] See on **97**.

lines 1–2 *dedicativus* is one of the words that Martha Kneale (**REF** p. 266) says shows Apuleius' dependence on Greek sources, rendering as it does *kataphatikos*. In line 2 is *reflexim*, which in AlexPA (Barnes) (See on **91A**) is related to κατὰ ἀνάκλασιν in **91D**.

Cassiodorus, *Institutions* 2.3.12 (p. 117.20–118.1 Mynors)

In section 12 Cassiodorus gives a straightforward account of syllogistic logic.

95B Alexander of Aphrodisias, *On Aristotle's* Prior Analytics 1.6 28a10 (*CAG* vol. 2.1 p. 95.25–32 Wallies) trans. AlexPA (Barnes) 168

AlexPA (Barnes) 168 n. 18.

Here Alexander refers only to some (*tines*) who held the view concerned, but it is clearly the same view as that discussed by Apuleius and attributed to Theophrastus in **95A**. Lynn Rose **116ff**. thinks that the case of Darapti was not all that he covered, but since all the other valid indirect moods in the second and third figures which he explored would be syllogisms already, Alexander did not comment on them.

Rose suggests that the whole of section 3–7 of Galen 11 could be a summary of Theophrastus' account. It is clear from *PA* 2.1 53a8–14 that Aristotle was thinking about all three figures in connection with the point that certain combinations of premises give more than one conclusion.

[95] Sullivan 152–3 espec. n. 60 deals with the order of moods in the third figure in Aristotle, Theophrastus, and Apuleius.

[96] Sullivan 102–3 says that Theophrastus was formally correct in having two syllogisms in Darapti. It could be that he was engaged in an independent exercise like some of those engaged in by Aristotle, e.g. at *PostA* 2. 5–7 discussed below at **110**.

96A Philoponus, *On Aristotle's* Prior Analytics 1.6 28b5 and 31 (*CAG* vol. 13.2 pp. 105.28–30 and 110.4–7 Wallies)

For Philoponus see on **71E**.

Here are two other cases of Theophrastus' interest in the order of items. He wanted to change Aristotle's order in the third figure because a) the reduction Of Disamis to the first figure involved the conversion of both one premise and the conclusion, so that he placed it fourth after Datisi which needed only the conversion of one premise, and b) he placed Ferison before Bocardo as fifth and sixth because Bocardo could not be proved directly, whereas Ferison could.

Since we are told (in **132**) that he made some changes in the order of the topics to facilitate memorising them, he may have been doing the same thing here.

Philoponus uses the word *tropos* for "mood" in line 3, in line with pseudo-Ammonius' usage, for which see on **106F**.

Patzig 74 suggests that the notion of degrees of imperfection lies behind Theophrastus' reordering of syllogisms.

96B Scholium on Aristotle's *Prior Analytics* 1.6 28a17 (cod. Paris. Gr. 2061 = p. 155.8–18 Brandis)

The scholiast names both Theophrastus and Alexander. Since Theophrastus is not named in our Alexander, the scholiast must have got his information either from another work of Alexander or from a separate source which names Theophrastus. See on **97**.

Alternatively Flannery xxi thinks this is based on the fact that Alexander in his logic is very sympathetic to Theophrastus.

The word *tropos* in line 3 must mean "method" of proof.

AlexPA 102.14–15, 102.32–103.2, 104.9–10, 106.16–19 trans. AlexPA (Barnes) 176, 177, 178, 180–1.

Alexander follows Theophrastus' order at 102.14–15, saying "which combination we (*hemeis*) have ranked fourth", and at 102.32–103.1 "we have ranked this combination third", giving Theophrastus' explanations without mentioning him by name. The same is true of his treatment of Bocardo and Ferison at 104.9–10 and 106.16–19.[97] In the last passage Alexander uses the verb "*ephamen*—we have said", which might refer to a work of his own.

[97] Bocheński 19 n. 47 notes that in Galen 10 (23.5–11 Kalbfleisch) Bocardo comes after Ferison.

97 Alexander of Aphrodisias, *On Aristotle's* Prior Analytics 1.32 46b40 (*CAG* vol. 2.1 p. 340.13–21 Wallies) trans. AlexPA (Mueller)[4] 23

It is uncertain whether the *logoi* ("arguments") here are the same as the *topoi* of DL's list (*The Reduction of Topics* **68** 18a = **1** 72), supported by the *Suda* (**2** 15) as one of Theophrastus' logical works. That work, if it was a separate one, may have contained reductions of Aristotle's topics to qualitative syllogisms, according to Barnes (1985) 135,[98] who refers to the fragment we give as **Appendix 2**. To these passages Graeser Fr. 19 adds Galen 11.7 which he discusses at 82–3. The text is doubtful. The passage might better be mentioned in connection with **93**. Bocheński 29–30 argues strongly against altering the ms readings.

See Sollenberger 46 no. 83, Solmsen (1968) 65 n. 1, and above on **68** no. 18.

This passage is evidence that Alexander knew two different works of Theophrastus, and it is possible that the order discussed in **96AB** was set out in the *Arguments reduced to Figures*, clearly an elementary work.

We should add a reference to Galen 16 (38,12ff. Kalbfleisch) which deals with a type of relational syllogism, like "Dio has half as much as Theon, Theon has half as much as Philo, therefore Dion has a quarter of what Philo has". Galen says that *hoi peri Aristotelen* (those around Aristotle) tried to treat them as categorical syllogisms, a procedure of which he clearly disapproves. This expression can hardly refer to Aristotle himself, but it could cover Theophrastus. Galen implies that there was some detailed work on this matter. AlexPA 244.13–27 discusses a similar case, but does not connect it with the Peripatetics.[99]

Appendix 1 Anonymous, *Analytics* col 1.1–17 (no. 3320 P.Oxy. vol. 47 p. 20 Philips)

Enough remains of this fragmentary passage to enable us to relate it to Aristotle's *PA* 1.33 47b29–39. The words *protasis adioristos* ("premise indeterminate") found here do not appear in that passage, although they are relevant.

---

[98] On this see Speca 41 n. 15.

[99] See Barnes, J. "'A third kind of Syllogism': Galen and the logic of relations" in R.W. Sharples (ed.) *Modern Thinkers and Ancient Thinkers*, London: UCL Press 1993 172–84 for a discussion of Galen's views here.

Using that passage one might expand lines 1–2 as:

> φθείρεσθαι αὔριον τὸν Σωκράτην ("Socrates die tomorrow"), and 4–5 as: ἡ μείζων πρότασις ἀδιόριστος οὖσα ὥσπερ καθόλου λαμβάνηται καὶ ("the major premise, being indeterminate, be taken as universal and")

In our passage the name is Socrates, in Aristotle's Mikkalos. Otherwise they are similar, and ours can be seen as written by someone who has Aristotle's *PA* before him.

Aristotle is writing about how to avoid errors in producing premises, here with the example:

> Mikkalus is musical Mikkalus
> Musical Mikkalus will perish tomorrow
> Therefore Mikkalus will perish tomorrow.

Aristotle's criticism of this is that it is not universally true that musical Mikkalus will perish tomorrow, and there could only be a syllogism if it were universally true. If we accept our reconstruction someone appears to be saying that the major premise is indeterminate, not universal. As we have seen on **82A**, Theophrastus took an interest in the indeterminate, which is some slight support for seeing his presence here.

Alexander on *PA* I 33 47b15–30 (AlexPA 351.24–353.6 trans. AlexPA (Mueller)[4] 37–9) has a number of possible interpretations of Aristotle's meaning, which suggests that the topic was much discussed.

The theme of a perishing musical being recurs in *On Generation and Corruption* 1.4 319b25–30 where Aristotle is discussing alteration and taking the case of a musical man perishing to be replaced by an unmusical one. The point here is the distinction between the man, who persists, and his properties, which can come and go.[100] To what extent these three passages are related we cannot tell.

Modal Logic

Becker 59 and 89–91. McCall. Patterson. Mignucci (1998).[101]

[100] See C.J.F. Williams' translation of *De generatione et Corruptione*, Oxford: Oxford University Press 1982 100–2 for an examination of this passage and the suggestion that there may be something wrong with the text.

[101] I limit myself to a few modern works. But for other logicians I add: K.J. Schmidt, "Eine modal prädikatenlogische Interpretation des modalen Syllogistik des Aristoteles' *Phronesis* 34 (1989) 80–106. F. Baddensiek, *Die Modallogik des Aristoteles in den Analytika Priora A,* Hildesheim, Zurich, New York: Georg Olms Verlag 1994 for a summary of recent studies. *CPSB* vol. 3 chap. 10 273–82 is helpful with background material.

Aristotle's modal syllogistic has been dismissed as a failure, although, as Barnes points out, he got it more or less right in the *DI*. Mignucci (1959) 67 regards *PA* 1.3 as one of the most unsatisfactory of the whole work, containing some inexactnesses and an ineliminable vicious circle. The Kneales 91 think it was a mistake to develop the theory of the modal syllogism, because syllogisms and modality do not mix. Either modal words modify predicates, and you can the use ordinary syllogisms, or modal words modify whole statements, and then all are affirmative and singular.[102] Benson Mates has said: "Unfortunately the reader is always left wondering what proportion of the difficulties (in Aristotle's modal logic) is due to the nature of the subject, what proportion to Aristotelian confusion, what to corruption of the text, and what to his own thickheadedness." Theophrastus was concerned with the framework of modal logic as Aristotle had set it up, and for him we may add to Mates' list the fragmentary nature of the evidence, and the fact that since he wrote there have been a number of further studies of Aristotle's modal logic, scarcely two of which are expressed in the same logical notation or give the same interpretations.[103]

I use the following notation:

> for syllogisms: BaA meaning "All Bs are A" etc.: for modal terms: N necessary, M possible and U assertoric, i.e. without modal qualification.

We must think of Theophrastus and Eudemus as being faced with Aristotle's ideas and working out their own positions. Aristotle proceeded as far as possible by using the same methods as he had used for categorical syllogisms. He used conversion to reduce syllogisms in the second and third figures to the perfect syllogisms in the first figure, and *reductio ad impossibile* for Baroco and Bocardo where conversion would not help. His progress was then affected by the fact that the negation of a necessary conclusion is a problematic one, and since he had not yet dealt with problematic premises he could not use *reductio*. Instead he used *ecthesis*. Patzig 127 is helpful here.

Lukasiewicz[2] 133 suggests that *PA* 1.3 and 8–22 are late and that Aristotle developed his modal theory late. If so, Theophrastus and Eudemus may well have been working alongside him.[104]

---

[102] For a reply to this see McCall 96.

[103] For more see Patterson 243–4 n. 12.

[104] N. Rescher, "Aristotle's Theory of Modal Syllogism and its Interpretations' in M. Bunge, *The Critical Approach to Science and Philosophy*, New York: Free Press of Glencoe 1960 152–77 notes a difference between the *PA* and the *PostA* about mixed syllogisms,

The following passages examine various cases in which Theophrastus is shown to have been concerned with problems in modal logic: **98** is on the nature of the assertoric or existential[105] proposition; **99** is on the problematic proposition; **100** is on necessity; **101** is on the relationship between necessity, actuality, and possibility; **102** is on the conversion of universal negative problematic propositions, and **103** on the complementary conversion of the same; **104** shows Theophrastus concerned with the logical ordering of syllogisms; finally **105–109** cover problems connected with syllogisms with mixed premises.

98A–G We have no material in Greek about Aristotle's view of the "assertoric" proposition, but a great deal from Arabic sources, Avicenna and Averroes.[106] They say that there were two different views about what Aristotle meant, those of Theophrastus, Eudemus and Themistius on the one hand, and of Alexander and later commentators on the other. Since there is nothing on this matter in our extant Greek Alexander, it is likely that they were able to use the *On the Differences between Aristotle and his Colleagues about Mixed Modes* to which he frequently refers in his *On the* PA. See on **106A**. Flannery 94 argues that this work was largely concerned with the nature of the "absolute" premise.

A word is needed about the language used here. In the translation Gutas has used "absolute" and "existential" as the best equivalents of the Arabic terms used, which themselves reflect the way Arabic logicians thought of the matter. But Aristotle himself is remarkably chary of using any term to describe this kind of proposition. Thus in dealing with arguments with necessary premises and conclusions he speaks of some propositions as being "not necessary", and rarely refers to any as ὑπάρχον or ὑπάρχουσα, whereas Alexander and later writers uniformly use the latter terms, for which we use "assertoric" or similar terms. Since Aristotle freely wrote of syllogisms in which these propositions played their part alongside the necessary and the problematic, he must have seen them as in some sense legitimate parts of a modal system, but it is not perhaps surprising that Theophrastus devoted some energy

---

and suggests that in the later one Aristotle may have been replying to arguments of Theophrastus.

[105] This is the most appropriate word in the Arabic context. See the discussion below.

[106] Pamphilus Montius (Panfilo Monti), who taught logic, medicine and philosophy in Padua and Bologna, and died in 1553, wrote a treatise on mixed modes in which he frequently quotes Averroes. But the references he gives are puzzling:

to working out his views. It would however be a mistake to see Theophrastus and Eudemus as just trying to work out what Aristotle himself meant. Rather, they were working alongside him and developing their own views.[107]

It is true that at **98B** and **E** Averroes says that this is what Theophrastus says that Aristotle meant in the *PA*. The rest of our passages make no such explicit claim, and it is likely that Theophrastus and Eudemus, having been given the modal framework, then tried themselves to work out a satisfactory account of the ideas involved.

It seems to be generally agreed that there were two parties who differed about the interpretation of the absolute or assertoric proposition, Theophrastus and Eudemus, followed by Themistius, on the one hand, and Alexander and the other later commentators on the other. One clear opinion (**98A**) attributed to Theophrastus is that an absolute proposition is one without the addition of either "necessary" or "possible". That is true, but superficial. More controversial is the claim (**98B**) that the absolute is like the genus of the necessary and the possible. In most of these passages one has to tease out the original views of Theophrastus (and Eudemus) from the interpretations of later writers.

98A Avicenna, *The Salvation,* Part One on Logic, Section on Absolute Premises (pp. 34.3–35.4, 36.6–12, 36.15–37.3 ed. Cairo 1913) (in Ibn-Sīnā, *an-Naǧāt*, Cairo 1331/1913).

For this work see Gutas 1988 112–4. *The Salvation* (1026 or 1027) was written for friends who wanted basic philosophical and scientific instruction without frills. The logical part was taken from his *The Shorter Summary on Logic* of about 1013–4. Clearly it was not all that simple and short. (PMH)

Avicenna is normally reticent about his sources. In this particular case, his mention of Themistius along with Theophrastus would indicate, as in numerous other occasions, that he drew his information from Themistius (see the discussion at **98B**).

---

a) many are not to our system of dividing Averroes, but it is possible to trace a systematic distortion in many cases, e.g. his Chap. 20 is our *Middle Commentary* Chap. 15. But some references are not traceable in our Latin text of Averroes even allowing for the distortion.

b) he sometimes says something different, e.g. in Chap. 17 he says that the *inventa* proposition is *quoddam potentiale.*

[107] See *PCSB* vol. 3 278 for more about whether the assertoric was a mode.

The example with the horses shows that by the term "transformation" (*naql*) of an absolute universal affirmation to an absolute universal negation Avicenna understood the negation of the subject of the premise. Furthermore, the same example indicates what Avicenna understood Theophrastus to have meant by absolute premises: absolute premises were premises that could be true if one disregarded the potential modality of the statement, that is, "without our knowing whether that is at the moment when B is predicated of it or at another time, and whether always or not always," as Avicenna says later on in the passage. Namely: "Every horse sleeps," i.e., at some time, and "Nothing that is a horse sleeps" if it happens that at a specific moment everything that is a horse is not sleeping. Aristotle mentions horses and sleeping in *PA* 31b8–10, but the specific combination in the example cited is not in Aristotle; presumably the example may have been coined by Theophrastus, though Avicenna does not say that.

By the non-transformability of the premises in some examples that Aristotle is supposed to have cited (lines 9–10), Avicenna most likely had in mind axiomatic statements which are always necessary, e.g., "The sum of parts equals the whole,"—an absolute premise by Avicenna's understanding of Theophrastus' theory since the consideration of time does not enter into it—where the negative transformation, "No sum of parts equals the whole," is invalid. (DG)

Here Avicenna states definitely that Theophrastus held that an absolute proposition was just one in which no modal word is mentioned. About Aristotle's view he is less certain, deducing what it was from examples in the text. Thus "Every horse sleeps" and "Nothing that is a horse sleeps" can both be true, so that the modal term "possibly" can be added. On the other hand, they say, Aristotle also gives examples of "absolute" propositions that are necessarily true, so that their negations cannot be true. Here Avicenna allows for two possibilities, i) that AaB and AeB may both be true, and ii) that both positive and negative premises are necessary, so that they cannot be true together, the view of Alexander.

In the next paragraph we are back with Theophrastus: Avicenna attributes to him a complicated account of the absolute proposition: that predication is either in the mind or in existence, and that it may be either at some time (only) or always.

The final paragraph confirms that account through a contrast with the view of Alexander, which is that if every B is A absolutely, A is predicated of a B only for a certain time. For Theophrastus the connection could be permanent, and therefore necessary, though the necessity would not be stated. (PMH)

G These are all from Averroes; as his sources he mentions Alexander, Farabi, and Themistius, and clearly makes much use of the last. He also appears to quote directly from Farabi, and knows a lot about Alexander. Flannery 56 n. 10 has a far-reaching study of these sources, and (93–4) argues convincingly that Averroes did not have direct access to Alexander's *On Mixed Premises*, but did have a full tradition to draw on. Certainly what he tells us is not from the Alexander we have.

E There exist Latin translations of the *Questions* by Renaissance scholars that add something to what we have in Arabic. Flannery 38 n. 14 has a detailed study of the MSS and the translations.

A problem here is the claim that Theophrastus said that the absolute premise was *as it were* a genus. But then it appears that the contents of that genus might be either two or three in number. Now, first, if there are three, the absolute, as one of the three, the necessary, the absolute, and the problematic, will be one of its own members. But, secondly, if there are only two the absolute cannot form part of the mixture of modes that play a prominent part in Aristotle's work, a point made in **98C**. A solution might be to distinguish between the absolute and the existential, with the former being the genus and the latter one of its members. But there is no positive evidence of such a distinction, and indeed in **98B** #s 114 and 115 we hear of an "absolute and existential proposition".[108]

Gutas thinks that **98B** contains the more precise formulation on the basis of what Averroes received from Themistius and Farabi, and because it is in agreement with Avicenna (**98A**).

Another problem is that Theophrastus seems in places to have stated that the absolute is only in the mind. That might fit the view that it is a genus only, and therefore abstract. See on **98D**.[109]

---

[108] Barnes writes: "Surely the thought is that if P, then either necessarily P or possibly P or, being necessarily thus and being possibly thus are the only two ways of being thus." This interpretation would fit in with the view that the absolute is only in the mind, but, in my view, leaves untouched the difficulty of using the absolute proposition as part of a syllogism. (PMH)

[109] I have toyed with the idea that while most of our material is closely connected with Aristotle's *PA*, and so with Theophrastus' work on that, Theophrastus might also have written independently on modal matters. But there is scant evidence for that. (PMH)

98B Averroes, *Questions* 4 (pp. 114.5–116.12 ʿAlawī)

D.M. Dunlop, 'Averroes (Ibn Rushd) on the Modality of Propositions,' *Islamic Studies* 1 (1962) 23–34. Rescher (1963) 91–105. C. Baffioni, 'Arist. An. Pr. A15, 34B7–18 nel commento di Averroè,' *Annali dell'Istituto Universitari Orientale di Napoli, Sez. Filol.-Lett.* I, 1979, 64–70. C. Ehrig-Eggert, 'Zur Analyse von Modalaussagungen bei Avicenna und Averroes,' *Zeitschrift der Deutschen Morgenländischen Gesellschaft*, Suppl. VI [XXII. *Deutscher Orientalistentag* vom 21. bis 25. März 1983 in Tübingen]: Ausgewählte Vorträge, W. Röllig, ed., Stuttgart 1985, pp. 195–199. Rosenberg & Manekin (1988) 96–97. Flannery, index. Gutas (1999)131–2.

Dunlop first edited this treatise, which was soon translated by Rescher. ʿAlawī then gave a second, better edition. Baffioni studied the particular Aristotelian passage mentioned in her title, while Chris Ehrig-Eggert studied the treatise by Averroes in the context of the latter's theory of modality. I have established the text on the basis of ʿAlawī's text and apparatus, using also the information provided by Dunlop and the Latin translation of Abraham de Balmes. Rescher's attempts to "correct" some of the readings have neither been taken into account nor entered in the apparatus criticus of the text volume.

The reference by Averroes in the second paragraph (line 4 of the Arabic text) to Themistius is to the latter's commentary, or paraphrase, of the *Prior Analytics* (Gutas (1999) 131–132). This paraphrase, for which cf. *RE* 1654,11–12, was known in Arabic, and Ibn-an-Nadīm reports that it consisted of three books: F 249.8–9 = Müller (1872) 27 = Peters (1968) 14; see also Lameer (1994) 6–7. Quotations from it appear to have survived in the Paris MS (Ar. 2345) of the Arabic Organon[110] and in a Hebrew translation from the Arabic (Rosenberg & Manekin (1988)).

The reference also in the second paragraph (line 4 of the text) to al-Fārābī must be to his great commentary on the *PA*, of which only the last quarter is extant.[111] Since the Aristotelian passage discussed by Averroes in this essay is 34b11–17, al-Fārābī's commentary on this passage, with the probable reference to Theophrastus, has not survived.

[110] See Peters (1968) 16 and note 6. The reference by Bocheński 21, n. 56, to Themistius' *PA* is a misprint for *PostA*.

[111] See Peters (1968) 16, who, however, was not aware of its survival. The extant portion has since been published in *Al-Manṭiqīyāt li-l-Fārābī* by M.T. Dāneš-Pajūh, ed., Qum: Maktabat al-Marʿašī, 1409/1988, vol. II, 261–553. It covers the text from 61b7 to the end. See especially Lameer (1994) 8–9.

In the argument by those who support the doctrine of Theophrastus against Alexander (line 20 of the Arabic text), the reference to rhetorical, dialectical, and demonstrative premises is to the late antique classification in Alexandria of the syllogistic propositions into five, corresponding to the last five books of the Organon: demonstrative (*PostA*), dialectical (*Topics*), rhetorical (*Rhetoric*), sophistical (*Sophistici Elenchi*), and poetic (*Poetics*). They range, in the order just cited, from being all true to all false.[112] The opponents of Alexander's view thus imply that such premises, since they should not be used in demonstrative, dialectical, and rhetorical syllogisms, can be used only in the two remaining, i.e. in sophistical and poetical syllogisms or arguments, which are, respectively, more false than true and all false.

The translation of the Aristotelian text (34b7–11) given in lines 26–29 is verbatim identical with the one extant in Arabic.[113] It was made by a certain Theodorus, revised by Hunayn ibn-Isḥāq, and used by the Baghdad Aristotelians. This would indicate that Averroes' exposition here is his own and not a mere transcription of the Arabic text of either one of his two sources, Themistius or al-Fārābī, which would have included the Aristotelian text presumably in a different Arabic rendering. Averroes thus appears to be recreating the arguments of the two parties he is discussing, and seeing that one of them refers to the text of the *PA*, he provides the pertinent quotation from his own copy of the Theodorus-Ḥunayn translation.

The examples that follow the citation of the Aristotelian text by Averroes are themselves the same as those given by Aristotle in what follows in the *PA* (34b11–17), except that they are recast in the form of a syllogism. (DG)

The course of the argument is such that I propose to set it out rather differently. Averroes says that he is drawing on Themistius and Farabi,[114] who agree that there are two views about the absolute premise, that of Theophrastus, Eudemus, and Themistius (the first party) on the one hand and that of Alexander and all the commentators except Themistius (the second party) on the other.

---

[112] See D. Gutas, 'Paul the Persian on the Classification of the Parts of Aristotle's Philosophy: A Milestone between Alexandria and Baġdād,' *Der Islam* 60 (1983) 231–267 at 241–243; repr. in Gutas (2000) no. IX.

[113] Edited by ʿA. Badawī, *Manṭiq Arisṭū*, repr. Kuwait 1980, vol. I, p. 183; edited by F. Ǧabr, *An-naṣṣ al-kāmil li-manṭiq Arisṭū*, Beirut: Dār al-Fikr al-Lubnānī, 1999, vol. I, p. 235.

[114] For a further reference to Farabi see **98D**.

For Theophrastus the absolute premise is itself without a modal term, but its matter is either necessary or possible,[115] that is, it is *as it were like* the genus of necessary and possible premises. This last remark is expressed tentatively, and we can only guess at what the original Greek may have been. Finally we are told that according to its matter *it is disposed* to have added to it one or other of the modes. This could be taken to mean that one can take a proposition as necessary or problematic according to its meaning. We may contrast this view with that of Iamblichus reported at **106E**, where an absolute or assertoric proposition may have a range between assertoric and necessary.

Alexander is said to have held that the absolute premise is a possible premise when it actually exists, i.e. in the present. This view is criticised by the party of Theophrastus on the grounds that such a premise would exist only by chance and for a very short time, and be useless for argument. They also quote from Aristotle *PA* 1.15 34b7–17 to show that he himself rejected such a view.[116] Note that these are not Theophrastus himself, but some presumably later followers.

It is similarly supporters of the doctrine of Alexander, i.e., if we take that claim literally, people later than he, who argue from the text, claiming that Aristotle was concerned only with premises in relation to what exists outside the mind, whereas Theophrastus held that absolute premises exist only in the mind.

The point that in an absolute proposition neither necessity nor possibility is stated is in agreement with what Avicenna says in **98A**, but that the absolute premise exists only in the mind is in conflict with the statement that predication may be of existence as well as in the mind (**98A** 11–14).

Flannery 94 n. 118 on line 5 notes that Elia's Latin translation refers to Herminus rather than Eudemus. While, in view of our other evidence, that is almost certainly wrong, he thinks it should not be forgotten. (PMH)

98C Averroes, *Questions* 8 (p. 177.4–13 ʿAlawī)

Flannery 96; Rosenberg & Manekin (1988) 96–97; Gutas (1999) 131–132.

---

[115] The notion of matter is difficult. See Flannery 98.

[116] Lines 26–9 is an accurate rendering of Aristotle, but in the following two 'quotations' what we have is a paraphrase only.

Averroes is doubtless referring here ("it became evident in another place") to his *Question* no. 4 (printed in the preceding fragment; *Question* no. 2 in the Latin translation), where he treated this subject in detail. His source is again Themistius's paraphrase, as discussed above. See **98B** and the bibliography cited there. (DG)

Averroes here clarifies for us that Theophrastus said that in the absolute premise the predicate belongs to the subject either possibly or necessarily. He goes on to say that this cannot be Aristotle's view, for on it the absolute premise is indeterminate and so cannot form part of standard syllogisms which have determinate conclusions. This information comes according to Averroes from "another place" which he does not identify, but see what Gutas says above. (PMH)

D Averroes, *Questions* 7 (pp. 156.11–157.1 ʿAlawī)

Rosenberg & Manekin (1988) 96–97. Flannery 94. Gutas (1999) 131–132.

Averroes' immediate source here is al-Fārābī's commentary, which itself may draw from Themistius. See **98B**. (DG)

Here we have a competing account of what is called throughout "the existential premise". Averroes refers to Farabi for the view that Theophrastus and his followers held that in such a premise "the predicate belongs to the subject absolutely, and that this [premise] subsumes the necessary, the possible, and the actually existent". In this context that ought to be an accurate account, if not an actual quotation, of Farabi's own words. Averroes himself appears to interpret this as meaning that what the existential premise subsumes is the necessary and "the actually existent which has the nature of the possible", but he admits that this is his own view of the implications of what they actually said. See on **99**. (PMH)

8E Averroes, *Questions* 6 (p. 148.5–7 ʿAlawī)

Flannery 97. Gutas (1999) 131–132.

In a different passage (**98B**, third paragraph), Averroes presented the doctrine of Theophrastus by saying that the absolute or existential proposition is like a genus of the other *two* modes, the necessary and the possible. This seems to be the more precise formulation, on the basis of what we assume him to have received from his sources (Themistius and al-Fārābī), especially since it basically agrees with Avicenna's understanding of the same sources (**98A**). In this passage, Averroes' formulation was possibly influenced by his interpretation of the Theophrastean doctrine as he explains it elsewhere, in **98C**, where he appears to regard the actually existent as a third category of the absolute existential proposition, in addition to the necessary and the possible. Averroes' own theory includes more subdivisions of both the possible and the absolute propositions (see Rescher *Studies* in *the History of Arabic Logic* 93 and the commentary on the following fragment). (DG)

Here Averroes states that Theophrastus held that Aristotle meant that the absolute was like a genus to the three modes. We have seen the difficulties of this account, and it seems unlikely either that Aristotle thought this or that Theophrastus believed it of him. (PMH)

98F Averroes, *Middle Commentary on Aristotle's* Prior Analytics 1.2 24b31–25a5 (pp. 143.18–144.2 Jéhamy)

Rescher *Studies in the History of Arabic Logic*. Rosenberg & Manekin (1988) 96–97. Gutas (1999)131–132.

In his *Question* no. 4 (partly cited above at **98B**), Averroes gives examples of his division of premises according to the parts of existence and of primary intuitions. His analysis of premises is as follows:

It is self-evident knowledge that the universal premise is the one whose predicate belongs to its entire subject and that it is of two sorts:

> The first is when the predicate belongs to the entire subject actually, and the second when it does not belong to the entire subject actually but possibly—I mean in future time. In some premises in which the predicate belongs to the entire subject actually, the predication is [valid] always, and these are necessary; in others, it is not [valid] always, and these are existential. The non-always [valid premises] are of two sorts: In one of them, the predicate belongs to the entire subject most of the time, and in the other, not most [of the time] but in lesser or equal [period of time]. The existential premises, then, are of these three sorts: [those which obtain] for the most, least, or mean [period of time]. And the same applies to the possible [premises]—namely, they are found to be those which [obtain] for the most, least, and equal [period of time].[117]

[117] Averroes, *Questions*, 117–118 ʿAlawī; cf. Rescher (1963) 98–99, 93–94. With these are to be compared the divisions of the necessary, possible, and existential given by Theophrastus himself as quoted by Themistius, in Rosenberg & Manekin (1988) 94–95.

This would seem to be Averroes' division of premises according to the parts of existence. As for the division according to primary knowledge or intuitions, it would seem to be referred to further down in the same *Question*. He says,

> The truly necessary premises are those concerning which the mind senses (*yašʿuru d̲-d̲ihnu bi-*) the essential relationship between the predicate and the subject; but as for when the mind does not sense this relationship, then [the premise] admits possibility, and when it admits possibility, it is existential, not necessary.[118]

It would seem that our primary knowledge establishes the necessary propositions, while considerations of existence and temporality further subdivide propositions into the various sorts of the possible and existential. (DG)

Averroes is here commenting on Aristotle, but in passing rejecting Theophrastus' account of the existential premise. At this point in the *PA* Aristotle is engaged in the preliminary classification of premises, and what Averroes is referring to is the *protasis tou huparchein* or assertoric proposition.

This work is also available translated into Latin from a Hebrew version of the original Arabic by Joannes Franciscus Burana of Verona. (PMH)

98G Averroes, *Middle Commentary on Aristotle's* Prior Analytics 1.15 34b7–8 (p. 200.15–22 Jéhamy)

Rosenberg & Manekin (1988) 96–97. Flannery 97. Gutas (1999) 131–132.

As in the other fragments collected in no. **98**, Averroes here is commenting on the section in Themistius's commentary or paraphrase which also mentioned Theophrastus (cf. Gutas (1999) 131). Averroes' word, "recommendation" (*waṣiyya*), refers to the term in the Arabic translation "ought to" (*yanbaġī*) which renders the Greek δεῖ (*dei*) in 34b7 of the Aristotelian passage (34b7–11). (DG)

This is more about Averroes' own views about the modal systems of Aristotle and of Theophrastus than about what they were.

---

[118] Averroes, *Questions*, 118–119 ʿAlawī; cf. Rescher (1963) 101.

The reference to primary intuitions may reflect Aristotle's *On the Soul* 3.8 432a12 where the meaning and importance of the words πρῶτα νοήματα ("primary concepts") are much disputed. (PMH)

99 Avicenna, *Twenty Difficult Questions*, Question on What Conclusion a Possible Minor and an Absolute Major Ought to Yield (pp. 96.22–97.4 Dāneš-Pajūh, in Mohaghegh and Izutsu)

In the *Prior Analytics* part ('The Syllogism') of his major philosophical summa, *The Cure* (*aš-Šifāʾ*, the *Sufficientia* of the Latins), Avicenna reviews five definitions of the possible in order to determine their validity. The first one reads, "The possible is that which is not necessary, but when it is assumed to exist, no impossibility results from it," which is precisely Aristotle's definition, couched by Avicenna in almost the same terms in which the Aristotelian text was available to him in the Arabic translation of Theodorus.[119] Avicenna thinks this is the correct definition. The second definition he quotes, which he later criticizes, reads, "The possible is that which does not exist, but when you assume it to exist, no impossibility results from it." This is identical with the definition of the impossible he gives in the *Twenty Difficult Questions*, where he attributes it to Theophrastus. The text published and translated at **99**, on the basis of the edition of the work by Dāneš-Pajūh in Mohaghegh and Izutsu, needs to be revised according to the correct text in the *Šifāʾ*. The last phrase of the text at **99** reads, "no inconsistency (or impossibility) is assumed; this should now be emended to read, "no impossibility results from it."[120]

Avicenna does not attribute any of these definitions to their authors in *The Cure*, in accordance with his practice, later in his life, to dispense with historical references in his work.[121] The *Twenty Difficult Questions*, addressed, the heading of the text in the manuscript tells us, "to Avi-

[119] Avicenna's text reads, *al-mumkin huwa llaḏī laysa bi-ḍarūrī, wa-matā furiḍa mawǧūdan, lam yaʿriḍ minhu muḥāl* (*Aš-Šifāʾ, al-Manṭiq, al-Qiyās*, Saʿīd Zāʾid, ed., Cairo 1964, p. 164.12–13). Theodorus's translation of the Aristotelian passsage (32a18–20) is very similar: *inna l-mumkina huwa llaḏī laysa bi-ḍṭirārī, wa-matā wuḍiʿa annahū mawǧūd, lam yaʿriḍ min ḏālika muḥāl* ('A. Badawī, *Manṭiq Arisṭū*, Kuwayt 1980, vol. I, p. 174).

[120] The last phrase of the text at **99**, *lam yufraḍ muḥāl*, should now be emended to read, *lam yaʿriḍ <minhu> muḥāl*. The mistakes in Dāneš-Pajūh's edition are palaeographically very common.

[121] See D. Gutas, *Avicenna and the Aristotelian Tradition*, Leiden 1988, 288–292.

cenna's contemporaries in order to test them", would appear to be an earlier work, compiled before *The Cure*; accordingly his reference to his commentary on the *Prior Analytics* would most likely be to his youthful summa of philosophy, *The Available and the Valid* [*of Theoretical Philosophy*] (*al-Ḥāṣil wa-l-maḥṣūl*), unfortunately not extant.

Avicenna does not name his source for the Theophrastus quotation in the *Twenty Difficult Questions*. The most likely candidate is again Themistius' commentary, though given the wide array of Greek sources available to Arabic scholars during the period of the Graeco-Arabic translation movement (ca. 770–1000), and to which Avicenna had access in the royal library of the Samanids, the dynasty at whose court he grew up in Bukhara, there can be no certainty in this regard. In *The Cure* Avicenna mentions three more invalid definitions of the possible, in addition to the two mentioned above.[122] One cannot but wonder who their authors might have been, and whether Avicenna found this information in the same source. In other words, the question is whether Avicenna compiled the list of the five definitions of the possible himself, on the basis of his knowledge of the Greek and Arabic philosophical literature, or he found it in one source which he basically followed. Further research into the history of Arabic modal logic may provide the answer.

As for Theophrastus's definition, it would appear that he kept Aristotle's formulation in his definition of the possible and merely either omitted the word ἀναγκαίου in *PA* 1 13 32a19, reading οὗ μὴ ὄντος, τεθέντος δ' ὑπάρχειν, or read instead, οὗ μὴ ὑπάρχοντος, τεθέντος δ' ὑπάρχειν. (DG)

This is the only passage directly concerned with Theophrastus' treatment of the possible premise.

As Gutas says, the definition of the possible given here is that of *PA* 1.13 32a19–20, and where our translation of the Arabic has "inconsistency" we should put "impossibility"; that is preceded by the clause "that which does not exist", which is puzzling, and difficult to reconcile with Avicenna's point b), that it is not a condition of a thing's being possible that it does not exist right now (i.e., presumably, at the time when it is possible.) This point is repeated obscurely at **100B** 1–2, which refers to

[122] These definitions are, no. 3, "the possible is that which is not necessary," without any additional qualifications; no. 4, "the possible is that which neither exists nor is necessary;" and no. 5, "the possible is that which is apt to exist and not to exist;" *Al-Qiyās* p. 164.14–16.

excluding the existent from the possible, but is not securely connected with Theophrastus.

It is tempting to solve the difficulty by assuming that somewhere in the long interval between Theophrastus and Avicenna the evidence has been corrupted, so that Theophrastus appeared to have substituted "what does not exist" for "what is not necessary" when he did not do so. But that theory would leave Theophrastus in agreement with Aristotle, and it looks as if Avicennna found some statement about his view which separated it from Aristotle's. We can go no further.

100A–D See Appendix 3 in AlexPA (Mueller)[2] 232–6, and Flannery 101–5. These passages are all on necessity, and most start from remarks by Aristotle himself. He recognised that there are various kinds of necessity. This is not a matter of linguistic ambiguity, but of different applications; there is an underlying core of meaning, but there are also complications which are difficult to sort out. **100A, 100B**, and **100D** are all from AlexPA but **100C** is from an Arabic translation of an attack by Alexander on Galen's *Essay on the Possible*, which contains an account of Theophrastus on the necessary.

To these passages we must now add the translation by the Hebrew scholar Todros Todrosi of excerpts from an Arabic translation of what appears to be the lost Commentary of Themistius on Aristotle's *Prior Analytics*, given as **100E** below. The English translation is in Rosenberg and Manekin 1988.

The central point is that Theophrastus recognised certain divisions of necessity more clearly than did Aristotle.

Aristotle himself discussed necessity in various places, without a system of cross-references. There are three chapters in the *DI*, 9, 12 and 13, in which he considers aspects of necessity and the interrelations of modal terms. Again in *PA* 1 3 and 8 he considers necessity within the framework of modal syllogistic. By the time of Alexander it had become a feature of interpreting Aristotle to compare various parts of his works and use them to clarify one another. It is clear that Alexander's teachers did that before him. But who began it?

Alexander (**100A** and **B**) says that Theophrastus (named alone) made the distinction, while the Arabic of **100C** refers to "the party of Theophrastus" which may stand for *hoi peri Theophraston*, which may in its turn mean just Theophrastus. Finally in **100D** Alexander refers to Aristotle's colleagues, normally Theophrastus and Eudemus, but nowhere is Eudemus named, whereas in **100D** there is an actual reference to

Theophrastus' *Prior Analytics*. It is possible, then, that Theophrastus alone, writing up his ideas, took the step of relating the discussion in DI. to modal syllogistic.

We have two separate quotations about necessity apparently from Theophrastus, the first (**100B**) from his *Prior Analytics* describing at least three cases of necessity, and the second, at **106B**, a straightforward argument about a modal syllogism, presumably also from his *Prior Analytics*. But they may well be entirely independent.

At this point we may raise the question whether Theophrastus and Eudemus made some kind of division of modal terms. Here in **100A** and **100D** the technical terms *diaphora* ("differentia") and *diairesis* ("classification") occur and there is an Arabic equivalent of differentia in **100C**. Further, in **100B** the word "third" implies some kind of enumeration, though not necessarily a *diaeresis.*[123]

Some problems are already there in Aristotle. Let us start with *PA* 1.15 34b7–8:

> We must take the "belonging to all" not limiting it by time (ὁρίσαντες κατὰ χρόνον) as to now or at this time, but absolutely (ἁπλῶς), for if we take the premise as referring to the present there will not be a (valid) syllogism. For possibly nothing prevents on occasion man belonging to every moving thing, for instance if nothing else were moving. And it is possible for moving to belong to every horse, but it is not possible for man to belong to any horse.[124]

Bocheński 81 assumes that Theophrastus, when he gave his examples, as in **102A**, did not know this passage of Aristotle. If he is right, it could be because the passage was not authentic, or because Aristotle wrote it in reply to Theophrastus. In favour of the latter alternative is the fact that Aristotle's example "Man belongs to every moving thing" is identical with the first example given by Theophrastus. Further, this passage follows one, *PA* 1 15 34b2–6, which has generally been rejected as spurious.

A Alexander of Aphrodisias, *On Aristotle's* Prior Analytics 1.3 25a29 (*CAG* vol. 2.1 p. 36.25–32 Wallies) trans. AlexPA (Barnes) 93.

---

[123] DL lists a *diaereseis* (*Divisions*) in two books by Theophrastus (**68** no. 15) but since that is unlikely to have survived into Alexander's time, and in **100B** we have an actual reference to the first book of Theophrastus' *PA*, we may ignore it.

[124] Ross's criticism of this example (340) is answered by Mignucci (1969) 327–8.

Alexander is commenting on the passage in which Aristotle considers the convertibility of necessary propositions, that NAaB converts to NBiA. Alexander thinks it appropriate to point out that Aristotle is here talking of propositions that are simply or absolutely necessary, and introduces, with an example, an alternative sense: "Man necessarily of every literate being (ἐστὶ ἄν ᾖ γραμματικός). This latter must surely mean "while there exists a literate being". This premise, he says, is not necessary without a qualification.[125] The point seems to be that "Man necessarily of every literate being" should convert to "Literate being belongs necessarily to some man", but because no man need be literate, and it could be that there were no literate beings in existence, even though there were plenty of men, there is something awry, and a further examination of necessity is called for. It is not clear why Alexander brings in Theophrastus. Possibly the latter made this point in connection with what Aristotle says here, but we do not know, though it is clear from **100B** that he did treat of these matters somewhere in his own *Prior Analytics* Book 1.

100B Alexander of Aphrodisias, *On Aristotle's* Prior Analytics 1.13 32a19 (*CAG* vol. 2.1 pp. 156.26–157.2 Wallies) trans. AlexPA (Mueller)[1] 93–4

Becker (1933). Bocheński 73–4. Barnes a) (1969), b) (1975a).

Aristotle has just finished his survey of the necessary and is turning to the problematic. Here he refers to the fact that the necessary is said to be possible homonymously.[126] Cf. 1.3 25a37 where he says that there are three senses of *endechesthai*, necessary, not necessary, and *to dunaton* (the possible). Here Alexander presents Theophrastus as being in agreement with Aristotle. He has a text from the first book of Theophrastus' *Prior Analytics* before him in which are listed at least three senses of necessary, though he does not tell us explicitly what the first two are. If we adopt Flannery's emendation (see below), Alexander himself gives us two senses. a) in itself, and b) of the existent. But Theophrastus' third resembles Alexander/Aristotle's second: the words used are not identical, but both refer to the necessity[127] of what is existent while it exists. The word

---

[125] AlexPA (Mueller)[2] in text and Appendix 3 has "on a condition". The expression *meta diorismou* is found only here and at **100D**.5.

[126] It must be 'possible' that is homonymous.

[127] The Greek words used here οὐχ οἷόν τε for "not possible' are rare in Aristotle,

huparchon here is particularly difficult. In Aristotle the context usually helps us to decide whether to take it as "belonging to" or "existent" without modal qualification. But the absence of modal qualification must be the primary point here. See **100C** for a further distinction.

A simple explanation might be that Theophrastus has been trawling through Aristotle and is here just referring to *DI* 9 19a23–7 where Aristotle is discussing the sea-battle, and says thar it is necessary that what is is when it is, and what is not is not when it is not. Against that is that Aristotle uses forms of the verb "to be", whereas Theophrastus uses *huparchein.*

Line 2: Flannery 103 n. 145 suggests as an alternative reading *kath' hauto*, which does away with "according to him" i.e. Aristotle, but that is not of great consequence. In any case the emendation is unnecessary. Barnes a)135 b)75–6 n. 54 suggests that Theophrastus and Eudemus were only repeating Aristotle's own objections, and he, like Bocheński 74, thinks that Theophrastus and Eudemus wanted a modal calculus of contingency for statements of the natural sciences.

In *LCM* 3 1978 90 Bob Sharples studies this and other related passages.

pseudo-Ammonius *On Aristotle's* PA 1.8 29b29 37.4

This is in a section headed On Necessary Syllogisms. It says that there is much *suggeneia* ("kinship") between the assertoric *huparchon* and the necessary: for the existent (*huparchon*) while it exists, is necessary. It goes on to explore other relationships between the assertoric and the necessary. The sentence "for the existent... necessary" is clearly based on Alexander (**100B**. 3–4) not on the sentence which Alexander quotes from Theophrastus in **100B**. 6.

C Alexander of Aphrodisias, *Refutation of Galen's Essay on the Possible* (cod. Escorial 798 Derenbourg f.59^{r})

N. Rescher & M.E. Marmura, *The Refutation by Alexander of Aphrodisias of Galen's* Treatise on the Theory of Motion, Islamabad: Islamic Research Institute, 1965; *DPA* I,136, no. 27; Rescher 1967. Sharples *LCM* 1978. Flannery 93, 99–101.

---

being found mainly in his *Politics.* Whether Theophrastus used the expression casually as a synonym for *endechesthai* and *dunaton* or deliberately with a shade of difference meaning "such as to' we cannot tell.

Lukasiewicz[2] discusses this passage at 151–2, but not helpfully. Rescher brings it in as evidence that the origins of Arabic temporal logic go back to Theophrastus. See n. 134 on **100C**.

This passage comes from the Arabic translation of the treatise by Alexander of Aphrodisias, *Refutation of Galen's Thesis on the Possible*, whose authorship and title are given as such at the beginning of the text *(Maqālat al-Iskandar al-Afrūdīsī fī r-radd 'alā qawl Ǧālīnūs fī l-mumkin)*. Only the title and the opening first two paragraphs of the Arabic translation of this text survive in the mutilated Escorial manuscript; the remaining pages of the text in the manuscript are lost, just as is the original Greek. A photograph of the surviving page of the Escorial manuscript is published by Rescher and Marmura (p. 153), together with an English translation of the entire extant page (pp. 67–71). (DG)

Alexander begins his treatise by quoting the words of Galen. The latter distinguished between two cases of existence, with the examples of the sun which is eternal, and man, who is passing, and said that Theophrastus incorrectly inferred from this distinction to a distinction among the senses of "necessary". (Barnes)

There is, as Sharples points out, a connexion here between time and modality. But it does not relate easily to the theory of the syllogism. For one thing, the sun is, for Aristotle, unique, and we cannot have a premise involving 'all suns'. We are left with the possibility that these points were made in a general discussion of the meaning of necessity, which could however, bear indirectly on the problem of the meaning of modal terms in syllogisms with mixed premises.[128]

Flannery 93 n. 113 thinks this may come from *On Mixed Premises*. (PMH)

100D Alexander of Aphrodisias, *On Aristotle's* Prior Analytics 1.10 30b37 (*CAG* vol. 2.1 p. 140.14–18 and 141.1–6 Wallies) trans. AlexPA (Mueller)[1] 75–6

Sharples *LCM* 1978. Flannery 75–8, 83–5.

In the second impression we have put "a contradictory pair expressed with regard to the future" for "the so-called contradiction about the future". (Barnes)

[128] Rescher has set out the complex developments found in some Arabic works, the core of which is the distinguishing and relating of what is necessary and what is always the case, and also making certain modalities dependent on the existence of the *subject* of a proposition. While speaking mainly of Stoic logic, he does refer to **100**. See also his 'The Theory of Modal Syllogistic in Medieval Arabic Philosophy' in *Studies in Modality*

Alexander is again allusive: the sense seems to be that the addition by Aristotle of the words "not however... absolutely necessary" indicates two things, a) that in genuine mixed syllogisms Aristotle meant that the conclusion would be of absolute necessity, and b) that he knew the distinction made by Theophrastus and Eudemus. In the gap between the two parts of **100D** (140.18–34) Alexander refers to certain interpreters of the topic of mixed premises. These are supposed to be defending Aristotle's position by saying that he means to conclude not the simple necessity but the qualified.[129] For they say that when animal belongs to all men necessarily and man to all that moves or walks, the conclusion is one of qualified necessity. For animal belongs to all walking or living things as long as the middle term belongs to them.

The translators, Ian Mueller and Josiah Gould, say nothing about these interpreters, who must precede or be contemporary with Alexander. As defenders of Aristotle they are presumably Peripatetics. Moraux (1984) 342–3 has argued for Sosigenes, one of the teachers of Alexander, in opposition to Bocheński 80 n. 259 who suggested Herminus, another of Alexander's teachers. Flannery supports Sosigenes.

On 141.1–6 Flannery argues that "the division" referred to here was spoken of by Sosigenes; he says that *hama de* and what follows parallels 140.16–17 (our line 3). The end of our passage refers to hypothetical necessity; Flannery points out that the same expression "τὸ ἐξ ὑποθέσεως ἀναγκαῖον" ("the hypothetically necessary") occurs both here in Alexander and in Philoponus, *On Aristotle's* PA 126.17 which he quotes and treats as important evidence for the lost *On Mixed Premises.* It refers to Aristotle's *DI*, but the relevant expression is not to be found in that work. Flannery concludes that it is Alexander's expression.[130] Our evidence connects it with propositions like "Every literate being is human" and the fact that human beings are transitory; "every literate being is

---

by N. Rescher (*American Philosophical Quarterly*) Monograph 8 Oxford: Basil Blackwell 1974, 52–3.

The problems involved, and some of the examples, are already there in the works of Aristotle and Theophrastus. Eclipses occur at *PostA* 1 8 75b33–5, 1 30 87b39–88a2 etc.; one can give a scientific account of an eclipse but it is not a thing that happens all the time. The account holds of the eclipse when it is happening. For more see on Philoponus at **100D** and **106D**.

[129] Flannery 76 translates *meta diorismou* as "with a limit", and finds that certain terms, e.g. moving, involve a time limit. He discusses limit further at 82–3.

[130] Flannery (75 n. 68) points out that it is used in a different sense by Pseudo-Ammonius *On Aristotle's* PA 39.29. See also Appendix 3 of AlexPA (Mueller)[2].

necessarily human" will be true only as long as literate beings exist. So the hypothesis concerned will be "If literate beings exist." That is not, however, stated explicitly anywhere. See also on **106D**.[131]

Flannery 83–4 discusses the purpose of Alexander's argument here, and thinks it is to show that Aristotle already had the relevant distinction in mind even though he does not introduce it in the *PA* until 1.10 30b33. But it remains unclear why Alexander mentions Aristotle's associates.

For good measure we can bring in a scholion on Boethius (Aristoteles Latinus III 1–4 p. 296 ad 25a1), for the source of which see on **113D**. That says: "Necessity is used in three ways, either when nothing is said of anything, (like "it is necessary for the sun to exist", for he who says this says one thing only) or when something is said of something (like "it is necessary for Socrates to breathe") or when an accident is said to inhere necessarily (like "it is necessary for Socrates to be sitting while he is sitting").[132] A similar claim is made in Boethius' *On Hypothetical Syllogisms* 839d ff. (Rescher 37).

There are both similarities and differences with Philoponus here, and also with Galen, reporting Theophrastus:

Philoponus	Boeth. Schol	Galen
1) the sun moves	the sun exists	the sun shines
2) Socrates is an animal	Socrates breathes	man is literate
3) Socrates is sitting while he is sitting	the sitter is sitting while he is sitting	

Boethius Hyp Syll.
1) God is immortal
2) Man has a heart
3) Socrates is sitting while he is sitting

The first cases involve the sun, which for Aristotle was eternal and necessarily existed and moved and shone. Boethius' scholiast has departed from the standard analysis and introduced the new idea of a single entity.

[131] Barnes points out that similar expressions with a similar meaning are to be found at *Physics* 2.8 199b 34 and *On the Parts of Animals* 1.1 639b25, but they are in a metaphysical context.

[132] The text goes on: *eisdem autem modis dicitur et inesse simpliciter. contingens vero praeter dictos modos dicitur cum id quod non est contingenter dicimus esse (ut contingenter omnem hominem esse album et nullum)*. ("*in the* same ways (something) is said simply to exist in (something). But besides the aforesaid ways 'contingently' is used when we say that what does not exist exists contingently (like 'contingently all men are white and none.' ")

If that is not an accident based in a misunderstanding, he was speaking of a necessary existential proposition, which is something different. Its source is uncertain. But "God is immortal", Boethius' own example, is more orthodox. The next case involves an entity that exists for a time only. Philoponus and the scholiast agree in having an individual subject, Socrates, whereas Galen and Boethius have a universal, man, but the point is still the same: what can be said of the subject is necessarily the case only while that subject exists. The third case involves a distinction between tne subject and the predicate of a proposition; here it is what is denoted by the predicate that can perish while the subject still exists, so that the proposition is only necessary while the predicate exists. Many of these examples are useless for syllogistic, and it is unlikely that Theophrastus would have discussed them in a syllogistic context. Even the Philoponus cases are not appropriate, though they are introduced in the context of the discussion of *PA* 1.9 30a15, after our **106D**. Philoponus refers to a *monobiblion* of Sosigenes, one of Alexander's teachers, for a view that takes up some remarks of Aristotle in the *DI*, beginning at 9 19a23. This interpretation goes far beyond what Aristotle says there.[133]

We are left with the suggestion that Theophrastus distinguished at least three types of necessity, but details are unclear. For more, see on **106DE**.

Extra text (**100E**)

Themistius, *Commentary(?) on Aristotle's* Prior Analytics Trans. from Hebrew trans. of Arabic trans. in Rosenberg and Manekin 91.

Theophrastus says that the necessary is said of three things the first and the most befitting the term is that which exists perpetually all the time, as in our saying "The heavens are ungenerated." It is proper that some things be above time, though not that they exceed time. For it is impossible that the First Being exceed time, but rather it is prior in nature to time inasmuch as it, *qua* first cause and prime mover, is the cause and originator of time. This is so because time is consequent upon the movement of the heavens as a whole, which exists by virtue of this

[133] Sharples refers to Ammonius *On Aristotle's* DI 153.13–154.2 for similar material.

mover. Since the mover is prior by nature to movement, it follows that it is prior [by nature][134] to time as well.

The second is that which does not exist perpetually, but rather exists of necessity as long as the subject exists, like the existence of soul in man and heat in fire.

The third is the existent thing when it exists, of which it is said that it exists of necessity, like the sitting of Reuven when he sits, or the walking of that which walks, when it walks.

Themistius' commentary on the *PA* has not survived. This passage has passed through two translations, from Greek into Arabic and from Arabic into Hebrew, and it is likely that it has undergone some alterations. Themistius himself is trustworthy, however, and the only obvious alteration is the use, in the Hebrew version, of the name "Reuven" where the Greek would have had, perhaps, "Socrates", as in the example above (Similarly the Arabic translators would have used "Zaid".). Further, I would be inclined to doubt whether Theophrastus used the example of soul existing in man. With these removals, the passage is in basic agreement with our other evidence. (PMH)

101 Boethius, *On Aristotle's* De interpretatione 12 21a34 *Second edition* 5.12 (pp. 387.27–389.20 Meiser)

Bocheński 87–91. Kieffer. Sharples.

We have placed this under modal syllogistic although it is closely connected with the *DI*, and may therefore be based on a passage in Theophrastus' *On Affirmation and Denial.* On the other hand there are differences, and the problems discussed are more complicated than those of the *DI.* It will be discussed further with the **107s**.

This long passage from Boethius' commentary on *DI* ch.12 attributes to Theophrastus a solution of a problem raised by the view that possibility" and "existing" (*esse*) follow (*sequitur*) necessity, and possibility is followed by neither *esse* nor necessity. But if the negative "not necessary" is equivalent to "possible", and "possible" follows "necessary", it looks as if "not necessary" follows "necessary". Theophrastus solved this by a study of the force of negation, saying that while "not neces-

[134] These square brackets enclose words added by the translators.

sary" is equivalent to "possible" the power of necessity is broken. He applied the same approach to "possible", for "not possible" is equivalent to "necessary". But they differ in that "necessary" is affirmative but "not possible" is negative.

This passage is preceded by a long discussion of the relations between *necesse esse, esse*, and *possibile esse* (and *contingens esse*) which is not directly related to Aristotle. It is something of a puzzle that Boethius alone reports all this, and we can only conjecture how he came upon it. It is true that he did pay great attention to the theory of negation and might therefore have picked out this passage where others did not. See on **72A** for a consideration of his sources; perhaps he got this passage from Porphyry. It is also possible that the original account was in Alexander's *On Mixed Premises*, which seems to have contained a considerable amount omitted, perhaps deliberately, from Alexander's *Commentary*.

Bocheński criticizes Theophrastus severely and speaks of a gross logical error. It is true that Theophrastus does not accept the solution given (more or less) by Aristotle at *DI* 13, where he distinguishes between "possible" as equivalent to "not impossible" (one-sided) and "possible" as equivalent to "not impossible and not necessary" (two-sided) (Ackrill's account). Graeser 74, like Barnes (1975a) 75–6 n. 54, suggests that Theophrastus here, as elsewhere, is using extra-logical considerations to make the difference. Bocheński goes as far as to suggest that even Aristotle was not entirely clear about the distinction he had introduced.

But what exactly is our passage about? In the long chapter 13 of the *DI* Aristotle is concerned only with relations between "possible" and "necessary", using a variety of Greek terms. But here Boethius adds *esse*. The nearest expression to that in Aristotle is *kat' energeian* ("in actuality"), but where Aristotle introduces the original puzzle about "not possible" etc. following necessity, there is no mention of *esse*.

The terms used, *akolouthein* (and *hepesthai*) in Greek and *sequi* in Latin, are rather difficult.[135] In Aristotle the Greek terms seem to mean something like "imply", but for Boethius could "*sequitur*" mean no more than "is inferior to"? We are reminded of the *deteriorem* rule introduced by Theophrastus (See **105–9**).

---

[135] Kieffer 77–82 first points out that at *Topics* 2.9 114b *akolouthia* is used of the relation of a property to its possessor, See Brandon in *Phronesis* 23 1978 173 for more references, and on **113C** for Hintikka. Kieffer goes on to a detailed study of terms used by Cicero and others in this area.

Elsewhere in this passage there is a group of metaphorical words which may be derived from Theophrastus' Greek: *confine* (line 27) does not mean "identical with", but rather "similar to"; *imitetur* and *consentiat* in line 37 are also cautious words, and the notion of "breaking the force of" is similar.

In line 12 we have had to use "possibly" for *contingere* because "contingent" would be odd here.[136]

We have printed in the apparatus a long sentence that is absent from our MSS but appears in the Basle edition, and we give the translation in a note to the main translation. This could be a genuine sentence, omitted by haplography. The view it contains, that what is possible is not yet the case, but will be in the future, is not found elsewhere, but at **99** Avicenna starts his account of Theophrastus' view with "The possible is that which does not exist"; see on **99** for more details.

102A–C and **103A–D** are on the conversion of universal negative problematic propositions, on which Theophrastus and Eudemus differed from Aristotle. Aristotle held that "A possibly belongs to no B" does not convert with a) "B possibly belongs to no A", but does convert with b) "A possibly belongs to all B". See Huby (2002) 102–5.[137] **102** is on the first point and **103** on the second. The material is mainly from Alexander, but there is interesting information from pseudo-Ammonius, not all of which is printed in FHS&G, because Theophrastus is not named. We therefore give extensive translations of some relevant portions. Menas' Anonymus is also available, though his value is less.

102A Alexander of Aphrodisias, *On Aristotle's* Prior Analytics 1.17 36b35 (*CAG* vol. 2.1 pp. 220.9–221.5 Wallies) trans. AlexPA (Mueller)[1] 137–8

AlexPA (Mueller)[1] 23–31.

In the second impression we have, following Barnes, substituted "they no longer say" for they are not also saying" in line 33.

---

[136] Martin, C.J. 'The Logic of Negation in Boethius', *Phronesis* 36 1991 277–304 is an extensive study of Boethius' very full treatment of negation, but he deals with modalities only at 301–2 and does not mention Theophrastus.

[137] Galen lists a work *On Problematic Premises* among his writings on the Peripatetics, but we know nothing about it. See on **73**.

Here we have two verbs in the singular form, *phesi* in line 2 and *deiknusin* in line 3, following a reference to Theophrastus and Eudemus. We cannot avoid the problem as we did at **90A** by supposing that Aristotle is the subject. Here Wallies has written the plural forms *phasi* (with the Aldine edition) and *deiknusin* (reaccented but unsupported). Mueller notes the reference back (line 1) to 41.21–4, which itself is a reference forward to this passage. There only Theophrastus is named, which might explain the singular here. But it is still awkward. See on **109A**.[138]

Let us analyse this passage:

1–8 set out the views of Theophrastus and Eudemus about the conversion of "possibly A belongs to no B".

9–16 is an anonymous defence of a position ascribed to Aristotle, which is not as Aristotle himself expresses it in *PA* 1.7 36b35–37a31, and seems to be by someone familiar with Theophrastus' and Eudemus' argument from separation. Note the use of temporal terms *hote* ("when") and *tote* ("then") for more on which see *PCSB* vol. 3 275–8.

In lines 10–11 there is a reference to what is "problematic according to the definition". This takes up Aristotle's own remarks at *PA* 1.14 33b22–4 and elsewhere, which refer back to his account of the problematic at 1.13 32a19–20 as "that which is not necessary, and when it is supposed to be the case, nothing for that reason will be impossible".[139]

In the argument here this is supposed to lead to the conclusion that B may be disjoined from A but not in the problematic mode, but how that is worked out is not clear. For more see below.

17–26 is Aristotle's own argument by *reductio ad impossibile* from *PA* 1.17 36b35–37a3.

26–32 gives Aristotle's second argument, from 37a4–9, but with examples of white things added. See Huby (2002) 104–5 for difficulties with Aristotle's argument.

32–6 sums up.

In line 20 *hêmin* ("by us") implies that Alexander is in agreement with Aristotle.

---

138 Barnes argues that the difficulty can be accounted for by the fact that MSS vary between singular and plural as expansions of the contraction φ and the absence of accents. But the problem occurs sufficiently often to make one wonder if there is something behind it, like a separate account about Theophrastus alone.

139 Mueller 19–21 discusses this strict contingency.

The first argument, that of Theophrastus and Eudemus, is a separation one, as in **90A**, but it also has a temporal element as in **106A**.

The source of what follows is unclear: it seems to be by someone familiar with Theophrastus' and Eudemus' argument from separation. I have suggested (Huby 2002) that it might still be by Aristotle himself in reply to Theophrastus and Eudemus. Barnes points out that it is not so much a defence of Aristotle's position as an attack on that of Theophrastus and Eudemus. Its central point is the idea that items can be separated but not *endechomenôs*, which we have rendered "in the problematic mode". On the other hand this does not sound like Aristotle, and he does not use *endechomenôs* at all, nor indeed *huparchontôs*. All we can say is that someone familiar with this argument produced this reply.[140]

We have added what follows because it refers (24–6) to Theophrastus' and Eudemus' view, in agreement with Aristotle, that the universal affirmative problematic does not convert. See on **103**.

pseudo-Ammonius, *On Aristotle's* PA 1.17 36b35 58.19–23 and 57.38–58.1

For pseudo-Ammonius see on **103C**. The items mentioned are part of a longer passage entitled: The Second Figure with both Premises Problematic (57.1–58.35), which forms a separate essay in its own right. It is of considerable interest, and I quote 57.22–58.23, which is related to the discussion of the convertibility of problematic negative propositions at *PA* 1.17 36b35–37a31. It falls into several sections, but I will give it as a whole, and then comment on individual points. I have decided however not to get involved in points of purely Aristotelian exegesis; there are many difficult passages which would repay investigation but are not related to Theophrastus. The main points for us are the mention of the Peripatetics in E) and the appearance of a separation argument at the end. I shall also say something about the Platonists. What we appear to have is an argument between some Platonists and some Peripatetics. We are told nothing about who they are,[141] and we cannot leap to the conclusion that Theophrastus is involved, but the possibility exists.

---

[140] Mueller 29–31 appears to attribute this approach to Alexander himself.

[141] Barnes regards this as an imaginary argument, pointing to "they would say" at 57.31 and throughout the argument. But there is other evidence for the existence of such Platonists. See on **103C**. Further, one must ask who produced these arguments if not some Platonists.

A) 57.22–31 "That therefore the universal problematic negative does not convert he now shows and sets out the proof in three ways. a) (36b36–37a9) If the A is possibly to no B, it is not necessary that the B also possibly to no A, for if this, the universal affirmative converts,[142] as I will show, but this is impossible: for white possibly belongs to all animal, but not also animal possibly belongs to all white things; for to some white things, like white lead, it necessarily does not belong. The conditional must be shown thus: If the A possibly belongs to no B, the B too possibly belongs to no A. Since what possibly to none, possibly also to all by way of the problematic conversion, both the A possibly to all the B, and the B possibly to all the A.

B) 57.31–33 "But the Platonists would say to this: not 'if the A possibly to no B, then also possibly to all in the case of the problematic as in the majority, but if at all, as by equality.' Aristotle would say: 'But the problematic stayed in its form (*eidei*)'.

C) (57.34–7) "The second proof (presumably 37a5–9) was by the same terms as those with which Aristotle showed that the universal negative did not convert. For white may possibly belong to no man, but it is not the case that man may possibly belong to all white things, for necessarily it does not apply to the swan (36b11) and white lead.

D) (57.37–58.1) "But the Platonists would say about the terms: 'What kind of white do you assume? If the absolute, not even the original problematic negative is that white possibly belongs to no man, for the white in a swan necessarily does not apply to the man, but if the white in man, possibly both white belongs to no man and man to nothing white.'

E) (58.1–4) "But the Peripatetics would say that the question had not been put rightly because the qualifications (*prosdiorismous*) had not been attached to the predicate but to the subject, and that as far as that goes the necessary universal affirmative converts."

F) (58.4–7) "But the Platonists would say that the actual terms had not been taken correctly; for they are on equality and not on the majority; and if we take them by the majority, the problematic universal negative converts: for it is possible for no man to be six-fingered and for nothing six-fingered to be a man." (Compare **103C**).

G) (58.7–13) "The third of Aristotle's attempts at proof[143] (*or* of the statement) that the problematic universal negative does not convert:

[142] That is, AaB would be equivalent to BaA.

[143] The MS has *lexeos*, but *deixeos* makes better sense and is an easy correction.

(1.16 37a9–30) if someone were to say thus: 'if the A possibly to no B, the B possibly to no A, for if not, necessarily to some: so that the A too converts with the B: but it was also possibly to none, which is impossible. The man who says this is arguing falsely: for to the 'possibly to none' is opposed not only 'necessarily to some' but also 'necessarily not to all'. And if that is supposed nothing impossible follows, for the particular negative does not convert."

H) (58.13–20) "But the Platonists would say: 'and how did Aristotle himself show in the three figures that both the problematic affirmatives convert with the particular affirmative? For he said: 'if the A possibly to all the B or to some, the B also possibly to some A; or if this is false, necessarily to none.' (more or less 25a40). For we too will say to you (*se*) that to the possibly to some is opposed not only necessity to none but also necessity to all. If this is assumed nothing impossible follows, since the A possibly to all B, even if the most impossible follows on if the A possibly to some B.' But the Peripatetics[144] would say that they had not used the criticized proof of Aristotle, but another like this: if the A has been separated from the B in a problematic mode, the B also has been separated from the A in the problematic mode, and for this reason the universal negative converts also in the problematic mode."

Comments

A) The main lines of the argument are in Aristotle, but the example of white lead is not there: he does refer to it as a case of white in *Nicomachean Ethics* 1.6 1096b23, but there the point is only that it has a similar *logos* ("account") to the white in snow.

B) So these Platonists accepted complementary conversion, i.e. the equivalence of AaB and AeB, in the problematic mode, but they then bring in a distinction between being problematic as equal, or as in the majority. This point reappears in F). To that Aristotle would (not did) say that the problematic stayed in its form. This is obscure, and there may be a textual error. Alternatively it has been suggested that it kept the "on equality" meaning.

C) Here again is the example of white lead, which is not in fact in Aristotle. Is something lost from our text of him?

---

[144] The original has "Platonists", but I adopt Barnes' suggestion of "Peripatetics".

D) The Platonists take up the case of different whites; they seem to be distinguishing between "Possibly white belongs to no man" which will not do because there are whites which necessarily do not belong to man" and "Possibly the white in man belongs to no man".

E) This is obscure: as we have seen on **84**, "qualification" might mean various things, at least either quantifier or added qualification, here presumably as "in man" in "the white in man". If we suppose the latter, the Peripatetics would be saying that a proposition like "Necessarily all whites in man belong to man" converts with "Necessarily all men belong to whites in man," but that cannot be the case. We may ask which is subject and which predicate here: presumably "white" is the predicate and "man" the subject, but how does that help when conversion is a possibility? Alternatively there is a textual problem: Wallies is responsible for τῷ κατηγορουμένῳ ("to the predicate") for the incomplete τὸ κατηγορουμε, of the MS and perhaps there is a deeper corruption.

F) Again the Platonists are wanting a form of possibility where one side is statistically more probable that the other.

H) The original text is obscure, but all becomes straightforward if we accept Barnes' suggestion. Then we have the statement that the Peripatetics used the separation argument which we have seen as favoured by Theophrastus and Eudemus.

AlexPA 41.21–3

Alexander covers Aristotle's treatment of conversion, and says that Theophrastus and Eudemus differed from Aristotle about the conversion of universal negative problematic propositions.

Scholium on Philoponus' *On Aristotle's* PA 1.3 25b14 (cod. Paris. Gr. 1917 = p. 150a8–10 Brandis)

In margin: "Theophrastus says that this too like the universal negatives converts with itself πλὴν ἐν ταῖς μίξεσιν ἐσται φάνερον". Is this last a separate comment by the scholiast? It presumably means "except in the mixtures (as) will be clear."

The comment is on Aristotle's sentence saying that when what is possible is what is more often than not and natural, the proposition is not convertible. It is not clear to what proposition the scholiast is referring.

See items B and F in pseudo-Ammonius, *On Aristotle's* PA 1.17 36b35 58.19–23 and 57.38–58.1 above.

102B Alexander of Aphrodisias, *On Aristotle's* Prior Analytics 1.17 37a9 (*CAG* vol. 2.1 p. 223.3–15 Wallies) trans. AlexPA (Mueller)[2] 141

This passage does not name Theophrastus, but there is no doubt that it refers to him and Eudemus.[145] It follows a long and difficult discussion by Alexander of Aristotle's 36b37–37a9, itself not easy to interpret. It contains a second argument about conversion:

To prove that MAeB converts with MBeA:
If it is false that MBeA
Then it is true that not-MBeA
Which is equivalent to NBiA
Which converts to NAiB.

But that is impossible because it conflicts with MAeB.

Alexander continues (223.14–224.11) with an account of Aristotle's reply to this argument at 37a14–20. He says that Aristotle's way of setting it out makes it obscure.

This passage is not in either Graeser or Repici.

102C Anonymous, *On Aristotle's* De interpretatione *and* Analytics (p. 100.1–22 Menas)

Mignucci (1965). (1998) 45. (1999) 25–6.

I am grateful to Hans Gottschalk for information about the Moscow manuscript. In that the item is called *Aporiai (Problems) of Theophrastus and Eudemus*. See also on **93**.

Mignucci (1965) 243–7 n. 37 argues with copious examples that the word *arnesis* supports the view that the work is late and from the school of Alexandria.

See Flannery 30 n. 74 and **93** and Mueller 209 n. 335.

[145] Mignucci (1965) 142–3 argues that Alexander was so meticulous that he would have named the colleagues if it had been known who they were. But Alexander is not always so precise. Thus at 140.32–141.3 he refers only to his 'colleagues' on a point which at 36.25–9 he attributes to Theophrastus, and at 156.29–157.2 even more specifically to Theophrastus in the first book of his *Prior Analytics*. Further, in the present case, on p. 223, he has referred as recently as 220.9 to Theophrastus and Eudemus, and there he is referring back to 41.27 where only Theophrastus is named. Lukasiewicz[2] 200 also doubts the report, because Aristotle had already refuted this proof (37a14–27) and Theophrastus and Eudemus must have known that. But if they had invented this account of the possible first this argument would not apply.

The first argument is described as expository (*ekthetikos*), which has raised difficulties. It is a separation argument like that given by Alexander in **102A**, but it sets out terms in the place of letters. Mignucci (1998) 45–6 wonders whether the term "expository" is the commentator's own expression.[146] In Mignucci (1999) 25 he thinks it an addition to material derived from Alexander. Barnes regards Graeser's emendation *t'auta* as absurd, and wants to change our text in two places, putting *ei* for *epei* in line 10, and keeping *tauta*. Mueller (1999b) 209 n. 335 accepts that an ecthetic argument is involved, but does not allow for the differing senses of *ekthetikos.*[147]

Here a similar argument to that of **102B** is actually attributed to Theophrastus and Eudemus, with illustrative terms "man" and "white" substituted for letters.

Lines 10–20 have some obscurities, which should probably be attributed to scribal errors:

Since it is false that M man e white	1
As it is true that M white e man	2
It will be true that not-M man e white	3

Now 1 and 3 seem to go together, with 2 intrusive. It is a repetition of the first half of the equivalence to be proved, i.e. M white e man is equivalent to M man e white. And in line 18 it is said that M white e man was assumed.

Much of this is close to Aristotle, but the last six lines (15–20) have no parallel in him. Nor are they close to Alexander. Since they contain the late word *arnesis*, and also *tropos* (modal word) for which see on **106F**, we should see them as late additions.

Graeser (Fr. 15) says nothing about his changes from Menas' text. Repici (fr. 21d) prints Graeser's text but translates that of Menas.

-D These are concerned with the conversion of problematic premises.

---

[146] Patzig 156–68 discusses *ecthesis* at length. He finds two senses in Aristotle, 1) proof of syllogism, 2) extrication of the three terms of an argument to test for validity

[147] One might speculate that there is a connection with Aristotle's *PA* 1.10 30b31–40. Aristotle says that he can demonstrate his point by setting out terms and has a syllogism with white and man and animal. He says: "For it is possible for animal to belong to nothing white. *Then* (the) man will not belong either to anything white, but not of necessity, for it is possible for a man to become white, but not as long as animal belongs to no white thing." The temporal aspect of the situation is clear here.

T-S. Lee (1984) 92–4 thinks the ideas about ecthesis of Theophrastus and Eudemus caught on, and he relates Theophrastus to extensional logic. See also Mueller. See Patterson 272 n. 14 for Peripatetic views on τὸ ὡς ἐπὶ τὸ πολύ.

103A Alexander of Aphrodisias, *On Aristotle's* Prior Analytics 1.13 32a29 (*CAG* vol. 2.1 pp. 158.24–5 and 159.6–15 Wallies) trans. AlexPA (Mueller)[2] 96.

Hintikka (1973) 31–4. Huby (2002) 103–5.

This is partly a commentary on a passage (32b4–22) of Aristotle which Becker suspected and Ross believed to be a note not connected with the rest of chs. 13–22. Aristotle distinguishes two senses of *endechesthai*, one equivalent to ὡς ἐπὶ τὸ πολύ, the other indefinite (*aoriston*), but says curiously that both are convertible. As Aristotle puts it, his view does not seem to work, and Theophrastus' objections seem to be valid. It should be noted that it is only here in Aristotle that the word *aoristos* occurs in a logical context, and that adds to suspicions of the passage.

Alexander comments here shortly, but refers forward to what we print as **102A**.

103B Alexander of Aphrodisias, *On Aristotle's* Prior Analytics 1.15 34b27 (*CAG* vol. 2.1 pp. 198.5–199.10 Wallies) trans. AlexPA (Mueller)[2] 109–11

Following Barnes we have in the second impression changed the translation of lines 6–7 to "just as would be the case if with this combination it were changed into "of necessity to some not". Barnes suggests no paragraph at line 12.

This is not in either Graeser or Repici, and its inclusion here calls for some defence. The passage is anyhow not easy, as Mueller brings out in his notes on it. Alexander is commenting on *PA* 1.15 34b27–35a2, which ends with Aristotle being dissatisfied with his own choice of terms to decide the nature of the conclusion of Celarent with assertoric major and problematic minor.

Alexander asks whether you can pass validly from the negation of "of necessity for some not" (NAoB) to "possibly for all" (MAaB). This would not be so if we could find examples with a) NAiB and b) not-NAoB, which would give not-M AaB. We cannot get beyond conjecture, but in what follows there are three examples which are Peripatetic, and could have been thought up by Theophrastus in response to Aristotle's admitted dissatisfaction.

Of his three examples Alexander, if we keep the text, rejects the first. He says that "No animal of necessity does not breathe" is *true* because a) some animals of necessity do breathe and b) some animals of necessity do not breathe. But that in fact contradicts "No animal of necessity does not breathe". In any case Alexander abandons this example, but presumably it had been suggested as possibly an appropriate example.[148] It does tie in with Peripatetic interests: Aristotle divided animals into those that breathe and those, like fishes, that do not, which is important for the theory of *pneuma*. The next examples are better, and are immediately followed (lines 39–40) by a reference to Theophrastus. Both are suitably Peripatetic, the rationality of man and god, and the motion of the rotating sphere, for which see e.g. **161A**.15.

103C pseudo-Ammonius *On Aristotle's* Prior Analytics 1.13 32a16 (*CAG* vol. 4.6 pp. 45.42–46.2 Wallies)

Huby (2002) 102–3.

The name pseudo-Ammonius is given to the supposed author of material found in the MS which has first a part of the genuine Ammonius, or at least a report of his teaching,[149] but that stops at f. 112ᵛ where there is an empty space and a comment by the scribe directing one forwards to what turns out to be by Philoponus. Wallies has added some scholia and "essays" from ff. 226ʳ–261ᵛ which he thinks are from the pages of a follower of Ammonius. Westerink 326 says that the essay on hypothetical syllogisms is a genuine work of Ammonius. See our comments on the apparatus to **102A**.

103C and **103D** are parts of a passage where pseudo-Ammonius is discussing modal syllogisms in the problematic mode. In it the views of some Platonists and some Peripatetics are considered. We cannot be sure who these are, but their arguments deserve consideration. Here Theophrastus and Eudemus object to complementary conversion of the problematic negative with the problematic affirmative on the grounds that it is not so with the interpretation of the problematic as involving

[148] Barnes queries the text and suggests either omitting *ouk* ("not") (in line 19), or adding *ouk* before *alêthês* ("true") in the same line.

[149] For a consideration of the relationship between Ammonius and his pupil Philoponus see K. Verrycken, "The development of Philoponus' thought and its chronology" in Sorabji (1990) 238–40.

being more probable than not. Barnes suggests "the possible/problematic" in the sense of "what holds for the most part" for τὸ ὡς ἐπὶ τὸ πολύ ἐνδεχέσθαι here and in **103D**.[150] He also urges "what is currently under discussion" for περὶ οὗ ὁ λόγος in line 4, as a stock phrase.

The Platonists referred to are puzzling. Flannery 65 n. 33 speculates about them, but with little result. See also on **106E**. We may speculate here, from the fact that in several places we come across references to them along with Theophrastus and Eudemus, (in **103CD, 106DE**), that they, or he, were contemporaries of the early Peripatetics. If so, the field is limited: Speusippus died in 339, which is too early, and Polemon was not interested in logic. That leaves Xenocrates (396–14) of whom we know only, for logic, that he wrote ten books of logical(?) problems (λύσεις τῶν περὶ τοὺς λόγους—DL 4.13), but all our evidence is late and may be misleading.

The passages from pseudo-Ammonius are included after **102A**.

For good measure I add the following from pseudo-Ammonius:
at 59.33–60.30 on *PA* 1.16 37a32 he says that Aristotle wants all the pairs in the second figure with both premises problematic not to give syllogisms. At 60.16 he quotes Alexander (230.13) inexactly, and goes on to say that the Platonists want as many pairs to give syllogisms as with the other modes, and oppose Aristotle's arguments as follows:

1) The universal problematic negative converts with itself, for if the A has been separated from the B, the B has been separated from the A. "For Aristotle admits the conversion of the problematic propositions, but not of the simple ones: the Platonists the opposite."
2) The Platonists have views about the *epi to polu* etc.
3) The definitions of the *endechomenon* are not of the *epi to polu* but of the equal.

At 65.5–15 there is a short account headed "Synopsis of the mixed (modes)" which sets out Aristotle's account in detail and adds "But his colleagues and the Platonists want the conclusion always to agree with the weaker (premise) in all the nine combinations."

See also the apparatus to **102A**.

103D pseudo-Ammonius *On Aristotle's* Prior Analytics 1.14 32b38 (*CAG* vol. 4.6 p. 49.7–12 Wallies)

This passage continues the thought of **103C**. It is rather allusive and not entirely clear: a) What are the ideas that Theophrastus and the others have about the first figure? b) Presumably they rejected the imperfect

[150] See Mario Mignucci 'Ὡς ἐπὶ τὸ πολύ et nécessaire dans la conception aristotélicienne de la science' in E. Berti (ed.) *Aristotle on Science The Posterior Analytics* Padua: Antenore 1981 178 n. 19.

but valid arguments in the first figure. c) What was included in the total of eight?

To take the last point first: Ross (285–6), in a footnote to his table of valid moods,[151] lists those that can be validated by complementary conversion of problematic premises. Most of these are mentioned by Aristotle, and it seems likely that the first four in the first figure, all of which are fully problematic, are the ones meant to be added to the original four assertorics. These would be the ones rejected by Theophrastus and the others; what exactly the ideas about the first figure were is still however not clear. Barnes writes: "Surely he *must* have meant to say 'So they will also have to change the first figure syllogisms'." He argues that since the text in the sole known MS. is wretchedly corrupt, perhaps we should write *kai* ("also") for *oude* ("not either") in line 2 to get that result.

104 Alexander of Aphrodisias, *On Aristotle's* Prior Analytics 1.8 30a9 (*CAG* vol. 2.1 pp. 123.18–24. Wallies) trans. AlexPA (Mueller)[1] 58

Flannery 29 n. 73. Mignucci (1998) 45. AlexPA (Mueller)[1] 11–13.

This passage has been misunderstood in the past.[152] It is evidence of Theophrastus' systematic approach. Aristotle himself saw that at this point in his system he could not prove a result by *reductio* involving the conversion of the conclusion, for that would lead to a syllogism with mixed premises, which he had not yet covered. He solved this by using *ecthesis*, but according to Alexander Theophrastus took the other route of leaving discussion of these cases till mixed modes had been dealt with. He may have decided not to use *ecthesis* because it seemed outside the system.

Mignucci (1999) 25 says that for Theophrastus, because Barbara NMM is valid one can deduce Baroco NNN, which Aristotle could not do and so had to use exposition.[153]

---

[151] Taken with some alterations from Becker (1933).

[152] See Bocheński 101 and Kneales 102. Alexander uses the same word, *hupertithetai* at 389.33 = **111E**.3 of Aristotle's postponing his treatment of hypothetical syllogisms. As far as we know Aristotle never produced such a work.

[153] Bocheński 76 constructed a proof by indirect reduction for Theophrastus, but it contains difficulties and there is no evidence for its having been used. Lukasiewicz and Patzig 157 suggested that Theophrastus and Eudemus did not understand *ecthesis*. The Kneales, in a private communication, wondered whether Theophrastus himself was uncertain about the general application of the *peiorem* rule, and they thought that it might have been taken up more firmly by his school.

Flannery says that in line 2 *peri toutôn* must refer only to apodeictic Baroco and Bocardo, and he is inclined to see this as part of an independent study of this topic and not part of a general commentary on the *PA*.

Barnes objects to the *d'* ("and") in line 6, because this sentence more or less repeats the previous one. The juxtaposition of the two is grammatically odd and the writing sloppy, but the meaning is clear enough.

105–9 One important difference between Aristotle and Theophrastus (and their followers) was about the mode of the conclusion in a syllogism with mixed premises. In a nutshell, Aristotle said that the mode of the conclusion followed that of the major premise, while Theophrastus said that it followed that of the weaker premise, the order being: necessary, assertoric, problematic. This is referred to as the *peiorem* or *deteriorem* rule, from the medieval Latin *sectetur partem conclusio deteriorem* (the conclusion should agree with the inferior part).[154]

105 Averroes, *Questions* 5 p. 123.5–8, 125.3–7, 127.7–18, 137,1–16

Rosenberg & Manekin (1988) 100–101, 103. Flannery 54. Gutas (1999) 131–132.

Averroes' source here is Themistius's paraphrase of the *PA*. See the texts in Rosenberg & Manekin (1988). (DG)

Flannery 53–108.

Most of this is an accurate account of what we know from other sources. But in 10–20 after a clear opening the argument is not entirely clear: it seems to make Theophrastus and Eudemus, in dealing with syllogisms with possible and necessary, and possible and existential (assertoric) premises say that Aristotle meant that the conclusions could only be possible. This is indeed in accordance with the vast majority of the cases he discusses. But the claim that Theophrastus and Eudemus said

[154] There is an exhaustive study of various attempts to interpret Aristotle in Patterson (1995) 87–123. See also Mignucci (1998) 44 n.15. W. Wieland 'Die Aristotelische Theorie der Notwendigkeitsschlüsse' *Phronesis* 11, 1966, 35–60 distinguishes Theophrastus' modal logic with its *peiorem* rule sharply from that of Aristotle. His general views are criticized in detail by Patterson 106–15.

that syllogisms with necessary and possible premises were perfect like the others is baffling. Presumably 'perfect' here is used only in the sense of 'genuine', not in Aristotle's more technical sense.

In 22–39 there is an analysis in terms of parts which we will find again in **106**; it is not very clear here. Averroes seems to be basing his claims on an actual text, and allowing for the vagaries of translation it is probably that given in **106B**.2–9. Hence I defer discussion to there. (PMH)

9 In the following sections there are many passages from Alexander, and taken together they are evidence both that Alexander knew Theophrastus' work and that the latter gave many detailed accounts of proofs on points on which he differed from Aristotle. (The position with Eudemus is less clear.) It is clear that Theophrastus was not content just to apply the *peiorem* rule.

–I Mignucci (1998) 44–65. Huby (2002) 93–7.

These are a miscellaneous group of items ranging from Alexander of Aphrodisias to Averroes, though the latter seems to have obtained his information from Themistius. The other authorities are Philoponus and pseudo-Ammonius.

Alexander (123.25) looks at Aristotle's treatment of mixed premises, beginning with necessary plus assertoric premises and saying that in the first figure if the major premise is necessary, so is the conclusion. This is immediately followed by **106A**. He then considers various interpretations of Aristotle by some unnamed people, followed by our **106C**. Finally several pages are devoted to discussion of Aristotle's arguments, mostly presented as his own criticism by Alexander, but with a final quotation from Theophrastus, our **106B**. Alexander lays stress on Theophrastus' "separation" argument, which is also found in **106D** (Philoponus) and **106E** (Pseudo-Ammonius), but does not feature in Averroes' passages at **106H** and **I**.

Alexander also gives the weak inductive, or analogical, argument (See n. 159), and some examples which are repeated in **106D** and **F**.

For related matters see *PCSB* vol. 3 278–80.

Mignucci records attempts by modern logicians to give accounts of Aristotle's point.

106A Alexander of Aphrodisias, *On Aristotle's* Prior Analytics 1.9 30a15 (*CAG* vol. 2.1 p. 124.8–30 Wallies) trans. AlexPA (Mueller)[1] 59

Flannery 53–108 esp. 98–9. Mignucci (1999). Huby 2002 92–7.

Barnes suggests that **106A** and **106B** should be transposed, because the latter gives a direct quotation from Theophrastus, the former only a smmary of his views.

Alexander wrote a study of the differences between Aristotle and his colleagues on syllogisms with premises of mixed modality. This is now lost, but much is known about it from his own references to it.[155] Flannery argues, however, from the evidence of the Latin translations of the *Quaesita* of Averroes that Averroes did not know that work.[156]

In this passage two points about Barbara NUN, i.e. Barbara with necessary major, assertoric minor, and necessary conclusion, which Theophrastus rejected, are attributed to Theophrastus and Eudemus,[157] the first that just as with affirmative and negative and with universal and particular the conclusion follows the weaker[158] premise, so with the modes. This is not a good point.[159] The second[160] is from the idea of separation or disjoining, which we have already met at **90A** etc.; the addition of modal terms does however add complications: a temporal aspect is introduced, as in **102A**, so that in the assertoric premise what is connected but not necessarily so may, or will, at some time be disjoined from it, and then the conclusion also will not be necessary.[161] A similar

---

[155] Alexander gives three versions of the title of this work. The two fuller are: ἐν τοῖς περὶ τῆς κατὰ τὰς μίξεις διαφορᾶς' Ἀριστοτέλους τε καὶ τῶν ἑταίρων αὐτοῦ γεγραμμένοις (*On Aristotle's* PA 125.30) and ἐν τῷ περὶ τῆς κατὰ τὰς μίξεις διαφωνίας Ἀριστοτέλους καὶ τῶν ἑταίρων αὐτοῦ (ibid. 249.38) See also 207.35, 213.26, 238.37. Flannery 68 suggests "disharmony" for διαφωνίας. Mueller 201 n. 239 suggests that Alexander thought these matters too difficult for students.

[156] We know nothing of the work *On Syllogisms with Mixed Premises* listed as one of his own writings by Galen. Presumably it used similar material to that used by Alexander, his contemporary. See on **73**.

[157] Here and only here Eudemus is named first. This may or may not be significant.

[158] Here in lines 5–6 Alexander speaks of the inferior and weaker (τῷ ἐλάττονι καὶ χείρονι) of the assumptions. "Inferior" is repeated in line 10. But ἐλάττονι can also mean "minor" and "weaker" is the more common term in this kind of context. Latin uses three terms, *inferior, peior* and *debilior.*

[159] Gisela Striker, 'Assertoric vs. Modal Syllogistic', *Ancient Philosophy* 14 1994 39–51 rejects this as a rule of deduction, being based merely on analogy.

[160] Barnes points out a mistranslation in line 11: for "it by this argument" read "this by argument".

[161] Mignucci 33 thinks that the sentence "A will not belong to [C] of necessity" could mean that A could possibly be disjoined from C. Mueller[1] 119 n. 32 finds Alexander's argument here unsatisfactory.

argument is given by Philoponus at **106D**.11–17 though in different words.

Finally, they give examples, of which the first involves animal, man, and moving. It is claimed that the conclusion, that animal belongs to every moving thing of necessity is false. Patterson 75–9 studies this example. Storrs McCall (1963 17) notes that Averroes in his *Middle Commentary* on *PA* 1.10 (trans. Burana) argues against this with: *ambulans non movetur ex necessitate, ex parte qua est homo, sed ex parte qua est ambulans... puta, ex necessitate, omnis homo movetur, ex parte qua est ambulans* ("a walking thing does not move of necessity from the fact that it is a man, but from the fact that it is a walking thing... consider, of necessity every men moves from the fact that he is a walking thing"). See Patterson 75–9 for a detailed study of this example. Two further examples are given (lines 20–5) preceded by *eti* ("again"). It has been suggested by Flannery (54 n. 6) that they may be Alexander's own examples. We cannot tell.

This passage of Alexander is immediately followed by "And this seems to be reasonable", and a short argument that seems to be Alexander's own. Flannery notes that in several places, of which this is one, Alexander appears to favour Theophrastus' views.

Flannery takes the matter further by suggesting that there may be traces of Theophrastus in the following section of Alexander (125.3–31). which ends with a reference to his own *On mixed premises* and an indication that it all comes from that work. It opens with a discussion of what Aristotle meant by saying that sometimes (*pote*) a syllogism is necessary when only one premise is necessary. (That seems to me not to be the way Theophrastus and Eudemus went about things: they did not pore over Aristotle's words.) Flannery also points out the similarities between this passage and what we learn of Herminus from pseudo-Ammonius *On Aristotle's* PA 39.32–40.2. Herminus was Alexander's teacher, and held the view that it was the matter (*hulê*) of the premises which determined the mode of the conclusion. Flannery brings in passages in Arabic from our **98** to support his suggestion that this was also Theophrastus' view. He also argues that the *oun* ("therefore") at 125.3 linked what followed with the preceding remarks about Theophrastus and Eudemus. But the immediately preceding lines are Alexander's own words endorsing to some extent their views, and the *oun* may well be Alexander's own. It is in any case a *men oun*, as Barnes points out. Further if an idea was attributed to Herminus it would not have been originated by one of his predecessors.

106B Alexander of Aphrodisias, *On Aristotle's* Prior Analytics 1.9 30a15 (*CAG* vol. 2.1 p. 132.23–34 Wallies) trans. AlexPA (Mueller)[1] 67

Flannery 53–108. Mignucci (1999) 33–6.

The syllogism involved has as premises AaB, necessarily BaC, i.e. Barbara UN?.

It is disappointing for those hoping to recover Theophrastus' actual words that in two passages from Alexander, **106A** and **106B**, we find the same argument expressed in different terms: in **106A** the verb is *apozeugnunai*, while in **106B** it is *chôrizein*. We might save the situation if we take it, as seems to be the case, that in **106B** Alexander is, unusually, quoting directly and at length, while in **106A** he is just summarising. Further, Mueller (n. 85) finds this passage obscure and suggests that that is due to Alexander's method of excerpting. We should also note that at **90A** it is possible that we have an actual quotation in which both words are used, though that is not presented as a quotation.

Mignucci (1999) 34 says that lines 7–9 contain a mysterious phrase which he interprets as meaning that Barbara NUN is valid only if the major premise is taken as meaning "A belongs necessarily to all that to which B belongs". For if C is one of the Bs, he argues, one can legitimately conclude that A belongs necessarily to C. He goes on to suggest that Theophrastus rejected this conclusion because it affected the conversion rule of necessary propositions, and he connects the problem with that of the analysis of identity. It would be too much to follow him into this area, but the ideas are well worth pursuing, and suggest that Theophrastus was breaking out of Aristotle's rigid system.

Barnes gives the gist of the matter thus: "Theophrastus means: 'If you interpret AaB as indicating that necessarily B holds of what A holds of, then in effect you turn AaB into NAaB'."

106C Alexander of Aphrodisias, *On Aristotle's* Prior Analytics 1.9 30a15 (*CAG* vol. 2.1 pp. 126.29–127.16 Wallies) trans. AlexPA (Mueller)[1] 61–2

Line 9: Mueller n. 47 argues that this claim, that Theophrastus and Eudemus accepted Bocardo MUM is problematic because it is equivalent to Barbara NUN which Theophrastus rejected. The claim is repeated at 248.19–30 (**107B**) See also Philoponus, *On Aristotle's* PA 124.1–4, a passage omitted in **106D**, which says that "Aristotle and everyone else" accepted Bocardo MUM. For lines 5 and 15–16 the translation is inadequate. Read: "If not, its contradictory, 'It is possible...' holds." (Barnes).

Alexander, having mentioned various other unidentified supporters of Aristotle and their arguments, now refers to more unidentified people who try[162] to support him with a *reductio* argument.[163] Mueller suggests that the words "try to show" indicate that Alexander was not happy with this argument, as he was not with other arguments—see lines 24–6, but for us the point is that he brings in Theophrastus and Eudemus as also holding that Bocardo NMM is valid.

Alexander, in what he appears to present as his own argument,[164] then turns to a second *reductio*, which uses Bocardo MUM, rather than Baroco; on this also Theophrastus and Eudemus are said to have agreed with Aristotle. Here a third figure syllogism produces the conclusion MAoB, and that is certainly incompatible with NAaB. Philoponus, *On Aristotle's* PA 123.26 gives this same argument as one used by supporters of Aristotle. That could, however, mean Alexander.

Earlier, at 126.9–23, Alexander has said that some of Aristotle's supporters have wanted to take the assertoric universal affirmative premise as true, not as a hypothesis, but as being the case and universal, so that it would not be proved false with any subject-matter. That would mean that it was necessarily true, and would exclude cases like "Man belongs to every moving thing", the kind of examples produced by Theophrastus and Eudemus. But Alexander then gives an objection which may be his own: "And yet what will they say when the minor is taken as no longer universal assertoric but as particular? For in this way, if the major is necessary he (Aristotle) says that the conclusion is necessary, and it is

---

[162] Barnes suggests *peirontai* for the ungrammatical participle *peiromenoi* in line 2.

[163] To prove NAaB.......1)
BaC........2)
NAaC.......3)

Use a *reductio*: negate 3)	MAoC
Add 1)	NAaB
To get	MBoC

which they claim is incompatible with 2) and is by a syllogism that Theophrastus and Eudemus would accept. But they would hardly have accepted that MBoC was incompatible with BaC. Barnes relates the argument to the views of those who would in effect turn AaB into NAaB, trying "to save Aristotle's system by an interpretation of non-modalised propositions which in effect makes them necessary".

no longer possible to say that the particular assertoric is not true—man of some walking thing" etc. That is:

NAaB
BiC
give NAiC

Could this last point have been made originally by Theophrastus and Eudemus? If so, as Barnes remarks, the supporters of Aristotle of 126.9 would be very early supporters, indeed contemporaries of Theophrastus, but we know so little of the various unidentified protagonists in these disputes that that is not impossible.

The references to Alexander are to places where he refers to his own *On Mixed Modes.*

106D and **E** are clearly related; Flannery has studied the connection between the longer passages in which they occur, namely Philoponus, *On Aristotle's* PA 123.12–126.29 and pseudo-Ammonius, *On Aristotle's* PA 38.20–41.7. He argues that both draw on Alexander's *On Mixed Modes* but notes that there are differences between them which he tries to evaluate. This is made extra difficult by the poor state of pseudo-Ammonius' text. Part of the Philoponus passage (126.8–29) is translated by Mueller (AlexPA (Mueller)[2]) 148–9 as part of his Appendix on Conditional Necessity.

106D Philoponus, *On Aristotle's* Prior Analytics 1.9 30a15 (*CAG* vol. 13.2 pp. 123.12–20 and 124.9–125.18 Wallies)

Flannery 53–108.

In the translation of line 53 replace *impossibile* with *absurdum.*

For Philoponus see on **71E**.

In a long passage Philoponus discusses the views of Aristotle, his supporters, and Theophrastus and Eudemus about mixed premises. For a later portion of the same section (126.17) see on **100D**, and for two short omitted sections (123.26 and 124.1–4) see on **106C**.

It may help to sketch Philoponus' strategy. He opens with the reckoning of the number of combinations of two premises involved. This leads up to our **106D** 1–10, which states that Aristotle and his colleagues differed about the mode of the conclusion in some cases. He then sets out the views of Aristotle and his supporters, first paraphrasing *PA* 1.9 30a15–23, with the introduction of the notion of parts which is not in Aristotle (123.21–6). He then adds two further proofs adduced by unnamed supporters of Aristotle's view, first a *reductio* denying the truth of the conclusion (123.27–124.4), and secondly the argument that if two assertoric premises give an assertoric conclusion, then if one of the premises is necessary the conclusion should not be the same, but assertoric (124.5–8).

There follows the rest of our **106D**. Here again there is a puzzle about singular and plural forms, for at lines 13,39, and 53 while Y has *phasi* BVt have *phesi*. In addition at 39 where we have put *legomen* ("we say") with BY, Vt have *lego* ("I say"). See on **102A**.

Philoponus now (**106D**.11–27) gives the views of Theophrastus and Eudemus, beginning with an attack on what he describes as "the Aristotelian proof", using the "separation" argument, which we have already met for categorical propositions (**90AB**), now applied to modal ones as by Alexander at **106A**.11–14. He enunciates more clearly than anywhere else that the meaning of necessity for Theophrastus and Eudemus is "what belongs to everything and always", whereas what is actual can "become apart from the thing". This looks like an actual quotation, introduced by "they say" (or "he says"), but we cannot be sure of that. Later (lines 24–5) it is said that the actual "is what actually belongs but can also not belong". That is at the end of a new argument about necessity, introduced by "they say", not given by Alexander. In it Philoponus speaks of what are really[165] parts, pseudo-Ammonius (**106E**.10) of the necessary parts of B. Only if C was really one of the parts or a necessary part of B would B necessarily belong to it. The nearest we can get to this in Aristotle is at *PA* 1.24 42a9–12: "something must be related to another as whole to part, for this has been shown earlier that when there is a valid syllogism it is necessary for some of the terms to be like this." It is not easy to find what it is to which Aristotle is referring, but at least we have the idea of parts and whole. That is also there very tersely at *PA* 1.1 24b26–7.[166] This could be applied to the problem of necessity. "A belongs to all B" means that every part of B is A. and that is all that[167] is needed for the categorical syllogism. But the situation is made more

[164] Flannery 75 n. 67 refers to Maier's view that it was Alexander's own theory, but rejects it.

[165] Barnes points out that the translation "real" is misleading in that the Greek word is an adverb, not a noun.

[166] Striker (1996) 216–7 examines Aristotle's rare but important remarks.

[167] We have seen above that Philoponus introduces the idea of parts in his paraphrase of Aristotle.

complicated by the introduction of modes. At *PA* 1.15 34b7–9 Aristotle says that we make syllogisms with universal premises that are always true, not true *kata chronon*, i.e. only at particular times.[168] There arises then a problem of distinguishing between what is just always true and what is necessarily true in the world as it is. Philoponus and pseudo-Ammonius give a solution by referring to different kinds of parts. This distinction is actually attributed to Aristotle's colleagues by both writers, but the different terms "necessary" and "real" are a problem, as is also the lack of discussion of this matter in Alexander. That might, however, be due to his having reserved it for his *On Mixed Modes*. There is an odd discussion at 129.2–20 where Alexander seems to be struggling to express his own ideas about necessary propositions, but there is no reason to connect them with Theophrastus.[169]

It is difficult anyhow to decide whether what are "really" parts or "necessary" parts are more appropriate terms here. Both involve distinguishing between some items under, say, B which are not only called B but have a further status which enables them to transfer their modality to, say, C. So we could have "All As are necessary parts of B; all Bs are C": but does that tell us that all As are necessary parts of C? Alternatively, do we get a better argument by speaking of what are "really" rather than "necessary" parts of B? I think not.

There is another problem in that the words of Philoponus and pseudo-Ammonius are not identical: Philoponus (**106D**.21–3) has: "<A> belongs of necessity to all <C>'s parts which are really parts of it, but not to those which are sometimes parts of it and sometimes not", but pseudo-Ammonius (**106E**.9–11) has: "it is not to every part of B... that A of necessity belongs". We can take the argument like this: it supposes that B, the middle term, has two kinds of parts, the necessary and the others. A belongs to the necessary parts only. And C, because it is only in fact related to B, is not of the necessary parts. (The translation is misleading here: there is nothing in the Greek corresponding to "one"; C itself could have parts and they be parts of B, but not necessary ones.) Philoponus, on the other hand, appears to be concerned with the conclusion, "A belongs to all C", but this does not make much sense. Averroes in **105** gives a straightforward account of this matter.

---

[168] See the introduction to **100A–D**. Barnes suggests that Aristotle is referring to syllogisms used in the sciences or to those with mixed premises.

[169] Aristotle himself tackles some related problems at *PostA* 1.8 75b21–35 and 1.6 75a1–11.

Bocheński 81 notes that at 124.9 Philoponus begins by repeating Alexander (our lines 11–35)[170] but then adds a *reductio* proof which Bocheński thinks is not by Theophrastus because it uses M_2 (two-sided possibility) and because Alexander does not mention it. To the former point one might reply that it could be part of an *ad hominem* argument as is also indicated by the remarks about the question of which conclusion is valid in lines 50–1.

Lines 32–5 contain a recipe for producing syllogisms that support the views of Theophrastus and Eudemus: the major term must be more universal (e.g. virtue), and the middle more particular (e.g. prudence) and there must be something, the minor term, that partakes of both (e.g. man). The origin of this recipe is not clear. It is in the context of arguments said to be put forward by Theophrastus and Eudemus, but is not actually attributed to them. It is repeated exactly in the marginal note to pseudo-Ammonius given by us at the end of **106E**.

At line 36 Philoponus turns to the *per impossibile* argument he has presented at 123.26–124.1 as being by some unnamed supporters of Aristotle. It goes: "We say, they say,[171] that with the premises being like this of necessity the A will belong to all the C. For if this is false the opposite will be true, 'It is possible that the A does not belong to all the C', so that again (*palin*) we may oppose mode to mode, quantity to quantity, quality to quality. Since then it is possible for the A not to belong to all the C, and it was supposed also that the B belonged to all that same C, it is possible then for the A not to belong to all the B.[172] But it was supposed also that (it belonged) of necessity to all (the B). The same therefore both possibly not to all and of necessity to all, which is impossible".

That is: To show that	NAaB	1)
	BaC	2)
Give	NAaC	3)
Negate 3)	MAoC	
Add 2)	BaC	
to get	MAoB,	

which conflicts with 1).

[170] Flannery (61 n. 22) notes that the examples at 28–32 are not identical with those given by Alexander in **106A**.18–25.

[171] This is another of the cases where the MSS have *phesi.* ("he says") which the editor has changed to *phasin* ("they say"). He is probably correct, but the point should be noticed.

[172] Flannery 60 n. 18 says that Philoponus calls attention to this mood which is not a standard Aristotelian mood and thinks it might come from a work on Aristotle's logic

A further argument (125.5–8) is that if you get an assertoric conclusion from two assertoric premises, you ought to get a different conclusion if one of the premises is necessary.

To this Aristotle's colleagues, i.e. Theophrastus and Eudemus, replied:

a) if two necessary premises give a necessary conclusion, one necessary and one assertoric ought to give something different.
b) such an argument should apply also where the premises are major assertoric, minor necessary, but Aristotle himself would have an assertoric conclusion here. "The same then to the same, both possibly not to all and necessarily to all, which is impossible".

They went on to produce arguments by *reductio* to show that both an assertoric and a problematic conclusion can be drawn from the same premises as those used by Aristotle's supporters: so the same premises can give AaC as well as NAaC: to show that

	NAaB	1)
	BaC	2)
give	AaC	3)
Negate 3)	AoC	
Add 2)	BaC	
To get	AoB	

which also conflicts with 1).

They went on to show that a problematic conclusion can be treated in the same way, so that by such *reductio* arguments one can show that NAaC, AaC. and MAaC are all valid conclusions from NAaB and BaC.

The argument of Aristotle's supporters appears with some differences in Alexander (**106C**.1–11) as by "some people". Alexander has "It is possible for the A not to belong to some (*tini*) C, whereas Philoponus has "It is possible for the A not to belong to all C" and only he has the remark about opposing mode to mode etc. In any case it is no more than a backing-up argument for the *peiorem* rule, and an attack on the particular case of NAaB BaC, NAaC.

If these arguments were indeed attacked by Theophrastus and Eudemus, they must have been put forward by their contemporaries. It is

which was not a commentary. It could have been reported in a work by Theophrastus himself. But then what of Eudemus?

to them that Theophrastus and Eudemus are said to reply by producing other syllogisms on the same lines. The whole account is doubtful: Barnes says that Eubulides did criticize some parts of Aristotle's syllogistic, but thinks it unlikely that he was interested in modal logic, and other contemporaries are hard to find. Hence the whole story may be a commentator's invention, but the possibility of its being true should not be ignored.

What follows in Philoponus is puzzling (125.18–126.29). He begins with what seems to be his own suggestion about why the combination discussed produces three valid conclusions, then replies to the difficulty about what, if two necessary premises give a necessary conclusion, a mixture of necessary and assertoric should give, saying that it should itself be mixed, which is ridiculous. Alternatively he appears to be saying that Aristotle's view that two mixed premises can also give an assertoric conclusion is equally problematic. He then asks at 126.7 if there is any way of helping Aristotle, and goes on to suggest one. In view of that we may conclude that the intervening passage is intended as an attack on Aristotle.

Philoponus *On Aristotle's* PA 123.21–5 and 123.25–124.1

The first is an account of Aristotle's argument, the second of an argument by his supporters which is quoted above.

E pseudo-Ammonius, *On Aristotle's* Prior Analytics 1.9 30a15 (*CAG* vol. 4.6 pp. 38.38–39.10 Wallies)

For pseudo-Ammonius see on **103C**.

This passage goes with **106F**. For a translation of the intervening passage 39.10–40.2 see Flannery 66–8. Pseudo-Ammonius' account is long-winded, dramatised and, in places, quite unacceptable, and the condition of the text is poor.

He makes Theophrastus argue that in the case of syllogisms with a necessary conclusion Aristotle does not give actual examples to prove his point, because he could not find any (**106F**.5–6). But he does report one good argument which fits in well with what we know otherwise of Theophrastus. He says that for Aristotle the conclusion of Darapti with one necessary and one assertoric premise will be necessary. We have already seen that Theophrastus, unlike Aristotle, recognised two forms of Darapti. See **95AB**. Given these we can then have

	NAaB	and	AaB
	CaB		NCaB
	NAiC		AiC
and, converting,	NCiA		CiA

and also

	CaB	and	NCaB
	NAaB		AaB
	AiC		NAiC
and	CiA	and	NCiA

The two on the left differ only in the order of the premises, but have different conclusions, and similarly on the right. In fact the distinction between major and minor premises is irrelevant in the third figure.

The passage **106E**.9–15 (39.5–10) is introduced by *phasi* ("they say"), which may take up Theophrastus, Eudemus and all the others mentioned before it, but may refer only to the later ones listed; in general it reports what we already have from Alexander (**106A** and **B**) and from Philoponus (**106D**). In addition, the examples given in the marginal note (movement, walking, man and virtue, prudence, man), and the remarks which follow, are almost identical with those given by Philoponus, (**106D**.28–35) though he does not have the diagram. There remains the question of the relationship between pseudo-Ammonius and Philoponus, the latter certainly being a pupil and reporter of the genuine Ammonius. I incline to the view that pseudo-Ammonius distorts Theophrastus and Eudemus, or whoever his source may be, even more than Philoponus does. The latter, as we have seen, refers to what are really (*kuriôs*) parts, but pseudo-Ammonius to the necessary parts. The marginal addition in pseudo-Ammonius is verbally almost identical with Philoponus. The diagram is of the same kind as that we give in **106F** from the following page of pseudo-Ammonius, but, as Barnes remarks, such diagrams were probably to be found in most of the commentators.

Iamblichus, named in line 5 as a follower of Aristotle on this point, (or a teacher of his) developed the view that both forms of syllogism were possible, with the example

> All men are necessarily animals
> All walking things are in fact men
> All walking things are necessarily animals

This leads to the view that the assertoric has a range, tending either to the necessary or to the assertoric, and according as it is one or the other so the conclusion will be assertoric or necessary.[173] See on **106F** for the reference.

Flannery 56 n. 9 regards lines 3–5 as evidence that Averroes (**98BC** etc.) used a hostile source for Alexander's *On Mixed Modes.* But all this says is that Themistius and Alexander were on different sides on one question.

In the apparatus we put *antropos* for *anthropos*, corrected in the second impression.

pseudo-Ammonius infra 65.14–15

This is at the end of a short section entitled *Synopsis of the Mixed (Modes).* It says "His colleagues and also the Platonists want the conclusion always to agree with the weaker (premise) in the nine mixtures."

F pseudo-Ammonius, *On Aristotle's* Prior Analytics 1.9 30a15 (*CAG* vol. 4.6 p. 40.2–16 and 37–40 Wallies)

Barnes has said, "the point is that it is arbitrary to reject the *peiorem* rule for modes while accepting it for quality and quantity." He therefore wants to keep line 2 as it is without *toutôn*, and translate "in the modes also". (Cf. **106E**.2–3). But one could see the following sentence as amplifying this one: the two cases mentioned are the random exceptions to the rule about modes.

A question arises about the meaning of the word *tropos* in pseudo-Ammonius. In line 2 we have translated it as "mood". We have then supplied "mood" in line 3, but there is no *tropos* there. In line 12 *tropos* must mean "mode". In fact pseudo-Ammonius uses *tropos* for "mood" frequently, as at 37.16, 43.19 and 20, and so on. But we need a word for referring to modalities, and the situation is surprisingly elusive.

Moods were firmly established in medieval logic and are treated as an integral part of Aristotelian logic. But the idea is not to be found in Aristotle[174] or in many of his Greek successors. Aristotle introduced the important notion of a syllogistic figure, and had an arrangement of

[173] This is in contrast with the view attributed to Theophrastus at **98B** that the assertoric can be necessary or problematic according to its matter.

[174] Patzig[2] 101 studies Aristotle's use of *tropos*, and suggests that such uses as do occur are without any fixed technical sense.

syllogisms under the three figures, but he and his successors managed perfectly well by speaking of, for example, "the third syllogism in the second figure". Not only Alexander in the second century AD, but even Philoponus in the fifth, mainly followed suit. Then pseudo-Ammonius used the word *tropos*, to cover what we would call both "mood" and "mode".

Let us look at an example:[175] In the opening pages of pseudo-Ammonius we find both mood and mode involved. A most revealing passage occurs at 38.1–3: "How with the assertoric did he demonstrate the fourth mood of the second and the fifth of the third through the impossible, not opposing mode to mode, quantity to quantity, quality to quality if indeed the assertoric is a mode. For it shows the fulfilled possible." In this context we cannot doubt that both mood and mode are covered by the same word *tropos*.[176] It is tempting to suppose that here the word has a more general meaning which covers both uses. But I have failed to find a suitable English word: "mood" stands for an actual arrangement of propositions, "mode" for a word or its meaning applied to a proposition.

We need not conclude, however, that *tropos* was not used in the sense of "mood" until this late date. I have suggested at **92** that it may have been found in educational works of elementary logic much earlier. Even so pseudo-Ammonius cannot be taken as quoting Theophrastus and Eudemus accurately here. It is anyhow suspicious that the material is so dramatized, and the references to *tropoi* with the meaning "moods", increase our suspicions.

The second objection is that Aristotle does not give examples to prove his case. Since that is a silly argument, for examples can only prove negatives, it is unlikely to be by Theophrastus. Barnes has suggested, however, that it might be the distortion of a "non-silly" argument. See the apparatus for a related remark.

It is not clear how much, if any, of what we have given is supposed to derive from Theophrastus and Eudemus. Barnes points out the change of style after the first two arguments at line 10, but in any case this

[175] In genuine Ammonius there is a cluster of cases on pp. 23 and 24 where it clearly refers to modes. It is concerned with *PA* 1.1 24b12, where *tropos* is not to be found. Also I have not found this use of *tropos* in the passages of Aratus (See on **93**) which I have seen.

[176] See Barnes, 'Ammonius and Adverbs' in Blumenthal and Robinson (1991) 145–63 for a survey of Ammonius' use of *tropos* as "mode" in a wide sense.

would only be a quotation by Pseudo-Ammonius from an intermediate source.

The next argument is that in the second figure the conclusion may, according to Aristotle's rule be either necessary or assertoric: the use of the word *tropos*, as we have seen, rules out the possibility of a direct quotation. Our final paragraph is expressed obscurely, but is supporting the possibility that one and the same pair of premises can have both a necessary and an assertoric conclusion.

The omitted portion of pseudo-Ammonius (lines 16–36) claims to give a solution provided to the writer by someone who got it from a *hupomnema* of Iamblichus. What it amounts to is a suggestion that an assertoric proposition can lean either towards the necessary or to the problematic, and so lead to a different conclusion.

We have said in the text volume that line 20 has an incomplete sentence. But Barnes points out that it is rather a rubric. What follows relates this theory to the views of the Platonists, for whom see on **106C**, and Theophrastus.

pseudo-Ammonius *On Aristotle's* PA 1.9 41.18–19

This is one of a series of short comments on Aristotle's text: 30a28: "again (*eti*) it is clear from the terms". "Now alone do you have terms, Aristotle", they say; "for why not also before, when you got a necessary conclusion by reasoning incorrectly (*paralogizomenos*)?"

We find here a similar dramatization to what we have printed. It does not occur in the surrounding notes, and we can only suppose that pseudo-Ammonius was drawing on an unusual, but hardly reliable, source claiming to report the views of Aristotle's colleagues.

G Philoponus, *On Aristotle's* Prior Analytics 1.9 30a32 (*CAG* vol. 13.2 p. 120.16–19 Wallies)

Aristotle is here very terse: "likewise too if the A B were negative; for it is the same proof". That is, if in Celarent we have AeB assertoric, but BaC necessary, the conclusion AeC will not be necessary. Aristotle's argument is by an example: movement, animal, man. Alexander's comment on this merely says that here Aristotle was in agreement with his colleagues, though not about when the major premise is necessary. We are left to suppose that his colleagues here, as elsewhere, used the *peiorem* rule.

For *horos* as premise see *PA* 1.24 41b6.

106H Averroes, *Middle Commentary on Aristotle's* Prior Analytics 1.9 30a33–30b6 (pp. 179.8–180.4, 180.18–20, 180.24–6 (Jéhamy)

Rosenberg & Manekin (1988) 100–101, 103. Flannery 54. Gutas (1999) 131–132.

Averroes' source here is Themistius's paraphrase of the *PA*: see the texts in Rosenberg & Manekin (1988). Averroes says something similar to the text printed in lines 5–23 in *Question* 5 (pp. 137.17–138.1 ʿAlawī), which should be added to the upper apparatus of parallel passages. (DG)

This like **106I** is based on Themistius. It deals with the mixture of necessary and assertoric premises, and says that Theophrastus followed the *peiorem* line. In lines 13–23 the words "They said" may introduce actual quotations from Theophrastus and Eudemus, as reported by Themistius. In them they distinguish between when the minor premise is necessary and when the major premise is necessary, but in each case claim that the conclusion will be assertoric. (PMH)

106I Averroes, *Questions 8* (p. 176.7–8, 176.12–177.4 ʿAlawī)

Rosenberg & Manekin (1988) 100–101, 103. Flannery 95. Gutas (1999) 131–132.

Averroes' source here is Themistius's paraphrase of the *Prior Analytics*; see the texts in Rosenberg & Manekin (1988). (DG)

Two examples are given in which a necessary and an assertoric premise give an assertoic conclusion. (PMH)

107AB These are concerned with syllogisms with a mixture of assertoric and problematic premises.

Here we may bring in the question of the consistency of Theophrastus' supposed system. It has been argued by the Kneales (102) that one cannot combine the three principles:

1) the *peiorem* rule
2) the universal validity of indirect reduction, i.e. proving the validity of a syllogism by taking the contradiction of the conclusion along with one of the premises and deducing a conclusion which conflicts with the other premise.

3) the validity of certain moods actually said to be recognised by Theophrastus, as here in **107**.

One case they use is:

Possibly all Bs are C
All As are B

Therefore Possibly all As are C. 1)

A *reductio* proof gives:
Negate 1) to get:
Necessarily some As are not C
All As are B

Therefore necessarily some Bs are not C.
But this violates the *peiorem* rule.

It can be argued however that Theophrastus was aware of this difficulty: our evidence comes, rather surprisingly, from Boethius in **101**. There we are told that Theophrastus was concerned with the effect of the negative on necessity, and said that the negative broke the force of necessity. If the *vis* of necessity is broken, we no longer have a problem, for the necessary premise is no longer stronger than the other one. This does not however take us all the way: it is concerned with "not necessarily", not with "necessarily not". But it does show Theophrastus concerned with the problems of how modal terms and negation are related.

A Alexander of Aphrodisias, *On Aristotle's* Prior Analytics 1.15 33b25 (*CAG* vol. 2.1 pp. 173.32–174.6 and 174.9–13 and 17–19 Wallies) trans. AlexPA (Mueller)[2] 83–4.

This is included in Repici fr. 30a.

Alexander repeats the points made by Aristotle about syllogisms in the first figure with a mixture of assertoric and problematic premises, in which he says that when the minor is problematic and the major assertoric the syllogisms are imperfect and the conclusion problematic only in the sense of being not necessary. *PA* 1.15 itself is long and rambling. We have omitted a) a passage in which Alexander reports Aristotle's statements about the perfection or its opposite of some syllogisms, and b) a passage about the differing senses of "problematic". Alexander then adds an obscure sentence about Theophrastus and Eudemus. What is the contrast implied by *mentoi* ("but") in line 12, to what does *tautas* ("these") in the same line refer, and what is the force of *kai* ("too") immediately preceding it? *Tautas* must surely refer to *protaseis* ("premises") back in line 3, and presumably the *mentoi* indicates some contrast with the views of Aristotle. *Kai* may indicate that Theophrastus like Aristotle called the premises "problematic", but in a different sense: the only likely account is that he held that in every case the conclusion was problematic according to the definition, i.e. neither necessary nor impossible. But the meaning is not clear, and perhaps there is something wrong with the text.

In line 6 Mueller, 188 n. 52, drops the second *to* ("the").

In lines 10–11 insert "problematic" between "to be" and "in accordance with".

107B Alexander of Aphrodisias, *On Aristotle's* Prior Analytics 1.21 39b32 (*CAG* vol. 2.1 p. 248.19–30 Wallies) trans. AlexPA (Mueller)[2] 174–5

AlexPA (Mueller)[2] 43–4.

This is included in Repici fr.30b.

Aristotle is concerned with syllogisms in the third figure with one premise assertoric and one problematic. As Ross says (366) the only case he actually deals with at 39b31–9 is Bocardo with problematic major, and that he deals with by *reductio*, using Barbara NUN which violates the *peiorem* rule and is therefore unacceptable to Theophrastus and Eudemus. So he has:

	MAoC	1)
	UBaC	2)
	MAoB	3)
From this we get		
Negating 3)	NAaB	4)
Add	UBaC	2)
To get	NAaC	

But this conflicts with 1); therefore 3) is false, therefore 1) and 2) imply 3). But Theophrastus cannot use Barbara NUN and instead for 1) substitutes UAoC to get:

UAoC
UBaC
UAoB

But this also gives MAoB, which is what we need. See on **106C**. Alexander seems to have before him Theophrastus' (only) alternative proof which involves changing the problematic premise into an assertoric. This leads to an assertoric conclusion.

Mueller deals at length with what he calls U-for-C substitution, which we should call U-for-M. It was used by Aristotle *PA* 1.5 34a25 (Mueller 39–40) where Mueller describes Alexander's treatment as unfortunate both as a piece of hermeneutics . . . and as logic. Again (44) Mueller says: "We do not know how to make sense out of Theophrastus' position—if he had one—but it seems clear that he was willing to use illegitimate U-for-C arguments."

Theophrastus was not the originator of this type of argument. Both he and Aristotle accepted it. This is odd if the position is really hopeless. One might approach the problem from the definition of the possible given by Aristotle at 1.13 32a18–21: "when (something) is not necessary, but is posited to exist (*huparchein*) nothing impossible will be because of this." And at 1.15 34a25–33 Aristotle himself sets out a U-for-M argument: if we suppose something unsupported by the evidence (*pseudos*) but not impossible the consequence (*to sumbainon*) which follows from the supposition will be unwarranted by the evidence but not impossible. See also Patterson (1995) 159–64 for an exhaustive study of Aristotle's position. He does not mention Theophrastus there.

In our passage Theophrastus is using a similar argument in a *reductio*, and explains carefully that the impossibility involved is not the consequence of the supposition but of the clash between Nxay and Mxoy.

Alexander's account is not entirely clear, and in the section preceding this he seems rather at sea.

B These are concerned with syllogisms with necessary major and assertoric minor premises. In each case Aristotle is arguing for a conclusion with which his colleagues, who are unnamed but must include Theophrastus and Eudemus, disagreed, because of the *peiorem* rule.

A Alexander of Aphrodisias, *On Aristotle's* Prior Analytics 1.21 39b32 (*CAG* vol. 2.1 pp. 247.39–248.5 Wallies) trans. AlexPA (Mueller)[2] 174.

Part of this is in Repici fr.30b, but it is not in Graeser.

This is followed shortly by **107B**. For its context see Flannery 71. Here Aristotle is dealing with syllogisms in the third figure with one problematic and one assertoric premise. At this point he tackles

UBaC 1]
MAoC 2]
MAoB 3]

This calls for a *reductio* proof:

NAaB (negation of 3))
UBaC 1)
to give NAaC
which is the negation of 2).

But Aristotle's colleagues, Theophrastus and Eudemus, objected to this, and said that the conclusion of the *reductio* proof would be assertoric (UAaC), and therefore not incompatible with 2). Instead they turned MAoC into AoC, as not being impossible.

Immediately following this, at 148.5–9, Alexander says that some people (*tines*) try to show by using the combination MAoC, UBaC with the conclusion MAoB, which has just been considered, that in the first figure combination with necessary universal major and assertoric minor the conclusion is necessary. As Flannery (71) says, Alexander clearly does not agree with these people, and we do not know who they are.

108B Alexander of Aphrodisias, *On Aristotle's* Prior Analytics 1.19 38a16 (*CAG* vol. 2.1 pp. 235.20–30 Wallies) trans. AlexPA (Mueller)[2] 156.

AlexPA (Mueller)[2] 215–6 nn. 413–9.

This is neither in Graeser nor in Repici.

In this section Aristotle is covering a variety of syllogisms in the second figure with one problematic and one necessary premise. At *PA* 1.19 38a16 he refers to a certain proof without giving it, and it is Alexander who actually gives the proof, that MAaC with NAeB give MBeC. This involves converting MCeB to MBeC, which is acceptable if by M we understand M_2 (two-sided possibility: CeB is neither necessary nor impossible). But, as Alexander says, opponents argue that NAeB with UBiC give only UAoC, which does not conflict with MAaC.

Barnes has pointed out that the proof (we should read *deixis*, not *deizis*) is itself a *reductio* which leads to an impossibility.

pseudo-Themistius, *On Aristotle's* PA p. 58.4–13

This work is not by Themistius, but is much later and consists largely of quotations from earlier writers. The editor, Wallies, mentions Sophonias (13 cent.) as a possible author. This section is part of a longer one taken from Alexander—57.33–58.13, between passages from Philoponus.

C These are a miscellaneous lot all concerned with syllogisms containing one necessary and one problematic premise, all of which, for Theophrastus, would have a problematic conclusion.

A Philoponus, *On Aristotle's* Prior Analytics 1.16 36b21 (*CAG* vol. 13.2 p. 205.13–27 Wallies)

AlexPA (Mueller)^2 202 n. 246.

For Philoponus see on **71E**.

Here in line 3 the editor, Wallies, has put *deiknuousi* ("they show") for the singular *deiknusin* of the MSS. As the subject is *hoi peri Theophraston* ("those associated with Theophrastus") this might be seen as a singular, but there is a similar peculiarity at **102A**; it is true that the problem at **90A** has probably been solved by supposing that Aristotle is the subject of the relevant verb, but that solution is not available here. See also on **106D**.

We are concerned here with Celarent with necessary major and problematic minor; Aristotle (*PA* 1.16 36a7–17) wanted an assertoric conclusion, as well as a problematic one, but Theophrastus argued for a problematic (only), i.e.

NAeB
MBaC
MAeC 1)

He argued that if 1) is false, not possibly AeC will be true, which amounts to necessarily AiC. We then get:

NAiC
MBaC
which give MBiA.

But this contradicts NAeB.

We have here a *reductio* proof of Theophrastus' claim. It does not however invalidate Aristotle's other claim, that an assertoric conclusion is also valid. Alexander gives none of this, and instead has a different *reductio* proof. See Mueller 202 n. 246 and 58 n. 74. It has been doubted if Philoponus is accurate.

At the end we have substituted "what is inferred" for "the inference" (Barnes) in the second impression.

109B Philoponus, *On Aristotle's* Prior Analytics 1.34 47b40–48a28 (*CAG* vol. 13.2 p. 328.1–9 Wallies)

This passage is in neither Graeser nor Repici

At *PA* 1.34 47b40–48a21 Aristotle is discussing the problem of inappropriate terms in a syllogism. He treats it as a matter of using abstract rather than concrete terms, e.g. "disease" rather than "ill", while Philoponus picks up Aristotle's term "states" (*hexeis*). It is difficult to tell how much of this relates to Theophrastus: minimally lines 3–4 only repeat the *peiorem* principle already enunciated in **107A**, but if all agree about the third figure then Theophrastus is at least by implication included.

In any case nothing affects Aristotle's main point about the care that must be taken in the selection of terms.

109C Alexander of Aphrodisias, *On Aristotle's* Prior Analytics 1.19 38a25 (*CAG* vol. 2.1 p. 236.11–14 Wallies) trans. AlexPA (Mueller)[2] 157.

AlexPA (Mueller)[2] 49–50, 216.

This passage is neither in Graeser nor in Repici.

Theophrastus is not named, but he is reasonably believed to be one of those who held that the conclusion in the syllogism under discussion, Camestres with problematic major and necessary minor, would be problematic, using the *peiorem* rule. Aristotle has sketched proofs of both Camestres MNM and Camestres MNU.[177] Alexander has been arguing that Aristotle's proof, by converting NAeC, which gives MAeB, only gives AeB and not MAeB because Mxey does not convert. He then brings in those who held a different view. It is not clear whether Alexander is actually drawing on Theophrastus and his like, or just taking

[177] Aristotle has argued that Cesare with necessary major and problematic minor leads to both an assertoric and a problematic conclusion. He then says that we can use the same proof with a negative premise, i.e. Camestres with problematic major and necessary minor, which gives a problematic conclusion. Aristotle's account is sketchy, but it is filled out by Alexander.

it that on the *peiorem* view this is what they must have held, and then producing an argument which they might have used.[178]

Barnes points out that *to prokeimenon* in line 2 is "what we are trying to prove", and that Theophastus and Eudemus were differing from Aristotle in what they thought the conclusion should be.

Prosleptic Syllogism

D Prior, A. *Formal Logic*² (1962) 121–5. Kneale, M. and W. 'Prosleptic Propositions and Arguments' in Stern et al. 1972, 189–207. Lejewski, C. 'On prosleptic Syllogisms' *Notre Dame Journal of Formal Logic* II 1961, 158–76. id. 'On Prosleptic Propositions' ibid. XVII 1976, 1–18.

Prosleptic arguments have attracted the attention of a number of modern logicians, who have carried the matter further than we can cover here. Prosleptic propositions are of the form "Of what A, of that also B', which implicitly brings in a third term beside A and B, and the name seems to come from the fact that the third term is taken in addition. The problem of dealing with such forms was invented by Aristotle, and set out in the difficult chapter *PA* 1.41. That opens with: "It is not the same, either in fact or in expression, to say that A belongs to all to which B applies, and that A belongs to everything to all of which B applies." The latter may be symbolised by (x) xaB ⊃ xaA, but the former is more difficult. Prior takes it as equivalent to "All Bs are A" (BaA), but the Kneales as xiB ⊃ xaA (If some xs are B, then all xs are A). We need not follow this matter and the further problems raised by Chap. 41, but should bear their existence in mind when studying Theophrastus. The theory of prosleptics was systematized by Theophrastus, and the Kneales have traced its history in the ancient world. Here I follow them.

The matter was also treated in *PA* 2.5–7 where, they say, Aristotle was engaged in "the curious intellectual exercise" of taking a normal syllogism and seeing what would happen if one took the conclusion and the converse, i.e. the proposition resulting from changing the order of the terms, of one of the premises and asking if one could prove the other premise in this way. Thus in Barbara we have:

---

[178] For a list of places where Alexander criticizes Aristotle's modal logic see Flannery 54–5 n. 7.

	AaB...	1)
	BaC	
Therefore	AaC	
Then we can take the conclusion		
	AaC	
add the converse of 1)	BaA	
and get	BaC,	
But with Celarent	AeB	1)
	BaC	2)
Therefore	AeC	
we can achieve a result by conversion of 2), but with		
	AeC	
	BeA (converse of 1))	
we get no conclusion.		

Aristotle asked how one could get a conclusion (BaC) here, and suggested that AeB could be seen as "B belongs to all of that to which A does not belong" (xeA) ⊃ (xaB), substituting C for x. Elsewhere he used another form, "B belongs to some of that to some of which A does not belong" (*PA* 2.5–6 58b9–10, 37–8, 59a28–9) i.e. (xoA) ⊃ (xiB).[179]

It is not clear from our sources how much of the later developments is due to Theophrastus, but we can be confident that he suggested the name "by proslepsis", and that he wrote something on the subject in his *On Affirmation*. He may well also have arranged the syllogisms in figures, but it is unlikely that he made a wide-ranging study of the matter.

110A Alexander of Aphrodisias, On Aristotle's Prior Analytics 1.41 49b27–30 (*CAG* vol. 2.1, p. 378.12–23 and 378.30–379.11 Wallies) trans. Kneales (1972) 190 and AlexPA (Mueller)[4] 70–1.

Lejewski (1961) 160–1 with nn. 10,11, 168–9 with nn. 29,30.

Alexander says a) that Theophrastus called these propositions *kata proslêpsin* (by proslepsis),[180] and b) that in the *On Affirmation* he showed that they differed only in expression from categoricals. If, as seems likely,

[179] See Prior 124–5 for a discussion of all this.

[180] The expression does not occur in Aristotle; at *PA* 2.5 58b9 there is, in some MSS, *dia proslepseos* (through proslepsis), but Ross (441) thinks that an interpolation from Philoponus. See on **110B**.

Theophrastus was following Aristotle's accounts, it could be that he said this about *DI* 7 17b12–16 where Aristotle says that "all" signifies not the universal but that it is taken universally, with the rejected example of "every man is every animal". See **84**.

On the face of it it is just a mistake to say that there is only a difference of expression between, say, "All As are Bs" and "A applies to all of that to all of which B applies". One solution is that Alexander and others got it wrong, i.e. misunderstood Theophrastus. Prior tries to go further, suggesting, as do the Kneales, that in some cases the two statements are equivalent, and that Theophrastus was misled by that fact. But Aristotle himself made the distinction at *PA* 1 41 49b14–32, and Theophrastus was perfectly capable of seeing the difficulty.[181] A cautious approach would be to say that a) he named and classified this type of proposition and b) by his choice of name he showed that he saw that a third term was involved.

Philoponus, *on Aristotle's* PA 2.5 58a21 (*CAG* vol. 13.2 pp. 417.8–12) trans. Kneales (1972) 192.

This is part of a section in which Philoponus comments in detail on *PA* 2.5 58a21–59a29. He uses the term *proslepsis* freely, but does not mention Theophrastus. See on **110B**.

AlexPA 264.2–5 = **112A**.51–5.

In the course of a long discussion of hypothetical syllogisms and the different words used by the Stoics and the Peripatetics Alexander touches on *proslambanei* and *proslepsis* and refers to the present usage.

0B Scholium on Aristotle's *Prior Analytics* 2.5 58a21 (pp. 189b43–190a5 and 190a18–25 Brandis) trans. Kneales (1972) 191

In line 16 for "syllogism" read "proposition" (Barnes).

This is on a passage of Aristotle in which some MSS. have the additional words (58b9) οὐκ ἔστι, δια προσλήψεως δ᾽ ἔστι ("it is not, but is by proslepsis"), which most editors reject, and may be from Philoponus 418.28, for which see the apparatus to **110A**.

Here there is mention of the figures, with examples of the first, second, and third figures. The Kneales (205) think it probable that it was Theophrastus who introduced the idea that prosleptics could be divided

[181] Lejewski (1961) 169 in a criticism of Bocheński introduces a new symbolic interpretation of what is involved.

into figures in the way that categoricals could, and they cite the fact that according to as good a witness as Alexander (**113B**), he divided totally hypothetical syllogisms in this way. Against this it might be argued that Alexander does not mention figures in connection with Theophrastus and prosleptics, but only much later authorities. Even they do not link the idea directly with Theophrastus. It may be significant that here Theophrastus is named both before (line 1) and after (line 7) the passage in which the figures are mentioned, but not in it.

Lejewski (1976) 1–3 starting from the Kneales, is also inclined to accept the claim that Theophrastus divided prosleptic propositions into figures. Most of Lejewski's paper however goes far beyond anything that Theophrastus could conceivably have held.

Scholium on Philoponus' Commentary on Aristotle's *Prior Analytics* 2.5 58a21 (cod. Paris. Suppl. Gr. 1156 = CAG vol. 13.2 p. xxxiii.19–24) trans. Kneales (1972) 190–1.

The scholiast says that Theophrastus calls such propositions *kata proslêpsin*. He repeats much of what we are told elsewhere, but without clear attribution to Theophrastus.

110C Galen, *Introduction to Logic* 19.1–5 (*BT* p. 47.18–48.21 Kalbfleisch) Kieffer 128–33.

The edition by Barnes and Frede has a number of improvements to the text, most of which do not affect the sense of what we give. Barnes points out that "commentaries" is unsuitable for *hupomnêmasi* in lines 6 and 19; "discussions" would be better. In line 9 we have corrected the translation to "what of tree, <of> plant also". Kieffer (4–5) discusses at some length the authenticity of this work. It is not named either in Galen's lists of his own works, or in any other of his voluminous writings, but that is not a fatal objection. It may be one of his latest works.

Galen does not name Theophrastus, but he does refer to the Peripatetics and what he says is in line with our other passages on this subject. In an Appendix (130–3) Kieffer discusses the usage of *neôteroi* ("newer", "younger"), and, in passing, cases of *palaioi* and *archaioteroi*, showing that there is a variety of uses, but concluding that Galen's use is neutral.

Galen uses the same forms of proposition as we find in other writers, but adds that we must understand in addition "is predicated" or "is said". This implies that in those writers these terms were not added, but at the same time it seems unimportant. He has clearly written more in his lost work *On Proof*, and there argued that these syllogisms are as it were condensed categoricals. The Kneales relate this to **110B**.7 where the scholiast says that these propositions contain a syllogism potentially. But there is no reason to ascribe all this to Theophrastus.

110D Scholium on *Aristotle's Prior Analytics* (CAG vol. 4.6 p. xii.3–10 Wallies) trans. Kneales (1972) 192.

This is a comment on part of a passage of pseudo-Ammonius (CAG vol. 4.6 p. 69) headed "On Prosleptic Syllogisms" which the Kneales think is intended as a summary of Ammonius' own views. It does not mention Theophrastus. They suggest that the scholiast puts forward the examples as true.

The Kneales (no. 8 p. 192) add the next sentence, about whether these are wholly syllogisms. But the verb is *phasi* ("they say"), not *phêsi* ("he says"), and it seems a late comment. They point out that the arguments in the second and third figures are incomplete, but that in both "Animal is of all men" must be supplied, as is clear from the diagrams.

Lejewski (1961) 158–9 studies the whole scholium.

pseudo-Ammonius, *On Aristotle's* Prior Analytics (*CAG* vol. 4.6 1) 67.27–8 and 2} 69.29–42 trans. Kneales (1972) 191.

1) says that the prosleptic proposition is equivalent in force to a syllogism, having two terms actually expressed and a third potentially. They can be reduced to the figures.

2) relates prosleptics first to categoricals, in that they can be reduced to the figures, and then to hypotheticals.

I add a final reference to a passage in Alexander's commentary on the *PA*: at 126.23–8, (trans. AlexPA (Mueller)[1] 104–5) just before our **106C**. Here he says that some unnamed supporters of Aristotle's view that NAaB, UBaC give NAaC bring in a prosleptic claim, that "A is said of all B" is equivalent to "A is said of all of that of all of which B is said". It is clear that these people are not Theophrastus and Eudemus, but see Mueller in AlexPA (Mueller)[1] 140 n. 287 for a suggestion that Theophrastus is involved.

The Kneales, in a private communication, have supported Bekker's suggestion that Aristotle's 32b25–37 should be excised, They say: "the whole thing looks like a note put in by someone who had thought more about prosleptics than Aristotle." That may show the influence of Theophrastus.

Hypothetical Syllogism

Barnes (1985). id. (2002) 114–23. Bobzien (2000) 87–137; id. (2002a) 358–94. id. (2002b) 57–72. Slomkowski 95–132. Fortenbaugh (2000) 65–79. Speca 114–23.

111–113 Aristotle speaks of syllogisms *ex hupotheseôs* (*PA* 1.23 41a37 and 29 45b15–21), but seems to cover a wide field by this expression and does not show a lot of interest in it. As **111** shows, Theophrastus and Eudemus developed his account, though even they did not go very far. We can probably discount the more extravagant claims made for Eudemus. Graeser 93 on his Fr 29 suggests that Philoponus was really concerned with the many works of the Stoics. The theme of Speca's book is that our evidence all dates from a time when people were confused about the origins of the terminology, which was first used by Theophrastus and Eudemus. The only area where we have full information about Theophrastus is that of "wholly hypothetical syllogisms", though here too the evidence of Alexander is sketchy.[182] Barnes (1985) 135–7 works out plausibly how Theophrastus could have treated the areas where we have no knowledge, but that is still conjecture.

There is clearly some similarity between the form of a hypothetical syllogism and the fundamentals of Stoic logic, but their relationship is still disputed. An early paper by Michael Frede[183] asks a pertinent question: why were the two at loggerheads on this matter, when to our way of thinking their views were easily compatible? Barnes (1985) 137–9 thinks they were not at loggerheads, but that the difference is that the Stoics developed a system whereas the Peripatetics proceeded piecemeal.[184] Slomkowski argues that the topics are cases of hypothetical syllogisms. One should, however, distinguish between what one might call natural methods of argument and the formalisation of such methods, and I do not find the latter in the *Topics*. Speca surveys the evidence we have, and (xi) says that his book "tells the history of a mistake".

111A–E are evidence that Theophrastus and Eudemus worked separately on hypothetical syllogisms, unlike the position with modal logic. It has been suggested that they discussed the subject with Aristotle and then wrote up material which included some of his ideas, but we do not know

[182] We can ask, but cannot answer, whether Theophrastus' view that these syllogisms can be related to the figures of the categorical syllogism was raised by him with Aristotle. Aristotle gives a firm denial that that can be done at *PA* 2.44 50b2.

[183] 'Stoic vs. Aristotelian Syllogistic' in Frede (1987) 99–124.

[184] In that paper Frede treats Alexander as the representative of Peripatetics after Aristotle, and ignores the fact that Theophrastus and Eudemus are reported to have developed the theory of hypothetical syllogisms in a way that is similar to that of the Stoics. Bobzien (2002a) takes a different route to identify the "ancients" with the early Peripatetics.

that for certain. We are however unusually well informed about where Theophrastus wrote about these matters: Alexander in **111E** mentions the *Analytics*, and in **112B** and **113B** the first book of the *Prior Analytics*, and Simplicius in **112C** does the same. Most scholars dismiss the evidence of Philoponus that Theophrastus wrote a great deal about hypothetical syllogisms, regarding it as inconsistent with our other evidence, but Barnes (1985) 127–8 makes a heroic attempt to harmonise all our material, though he concludes only that there is "no sufficient reason for disbelieving" Philoponus.

The Stoics and Peripatetics differed widely on logic; in addition they used different terms for the elements of hypothetical syllogisms, and I give a table:[185]

Source	Ancients (Peripatetics)	More recent (Stoics)
Galen	Hypothetical by connection (ὑποθετικός κατὰ συνέχειαν)	Conditional (συνημμένον)
Philoponus	Conditional (συνημμένον)[186]	
	Hypothetical through connection (διὰ συνεχοῦς ὑποθετικός)	Conditional tropic διὰ τροπικοῦ
	Antecedent (ἡγούμενον)	Antecedent (ἡγούμενον)
	Consequent (ἑπόμενον)	Consequent (ἑπόμενον)
		'Legon' (λήγων)
Galen	Separative (διαιρετικός)	Disjunctive (διεζευγμένον)
	Hypothetical by separation (ὑποθετικοί κατὰ διαίρεσιν)	
Philoponus	Substituted assumption (μετάληψις)	Additional assumption (πρόσληψις)

[185] Slomkowski 108 considers the origins of this terminology. Speca 35–65 devotes a chapter to trying to show that the commentators from Alexander onwards had conused ideas about how Peripatetic and Stoic logic were related.

[186] Frede 81 n. 18 says that by the time of Philoponus *sunêmmenon* was a part of common usage.

Alexander (**112A**) and 324.16–19	Assumed in changed form (μεταλαμβανόμενον)	Assumed in addition (προσλαμβανόμενον)
Alexander (**112B**)	Assumed in addition (of Theophrastus) (προσλαμβανόμενον)	
Philoponus	Conclusion (συμπέρασμα)	Epiphora[187]
Alexander	By analogy (κατὰ ἀναλογίαν) (of Theophrastus) in threes (διὰ τριῶν)	
Philoponus	Totally (δι᾽ ὅλων / δι ὅλου) hypothetical (ὑποθετικός)	

111A Boethius *On Hypothetical Syllogisms* 1.1.3–4 (p. 206.4–17 Obertello)

Dürr. Obertello. Speca 35, 67–80.[188]

There is some uncertainty about Boethius' sources. Dürr 5–9 dismisses the suggestion that Stoic logic was the basis of his work. In particular he omits the third indemonstrable inference scheme—not both the first and the second, but the first, therefore not the second. Obertello (85–7) thinks that he did have a Stoic source, but the arguments are not very clear. Speca 70–6 studies both Dürr and Obertello in detail, and finds both wanting to some extent.

We know of several earlier Latin writers on hypothetical syllogisms, Cicero's *Topics*, which alone survives, the *On Hypothetical Syllogisms* of Marius Victorinus, and the work of Tullius Marcellus, who wrote seven books of which the fourth and fifth were on hypothetical syllogisms;[189] to these Obertello adds Martianus Capella, whose *On the Wedding of Philology and Mercury* has as its fourth part a short work *On the Art of Dialectic*, and that has a section on hypothetical syllogisms derived from Stoic sources.

[187] For other examples of this term see Repici 142 n. 194.

[188] The title of Speca's work is misleading. It is as much about the Peripatetics as about the Stoics.

[189] We know of Tullius Marcellus only from Cassiodorus, and cannot date him (See Prantl I 664), but his work must have survived to be known by Cassiodorus, a contemporary of Boethius, who died at an advanced age in 583. Cassiodorus *Institutiones* 2 3 13 p. 119 Mynors) seems to have used him and Marius Victorinus.

But not only does Boethius deny that he used Latin writers: he does not have the seventh, or the third and sixth inference types which correspond to Chrysippus' third. All the Latin writers have seven types, closely connected with the five indemonstrables of Chrysippus.[190]

Contemporary with Boethius is a scholion ascribed to pseudo-Ammonius, for whom see on **103C**, which mentions five schemata, the same as those of Chrysippus, and therefore not Boethius' source.[191]

Boethius' use of the terms second and third figure agree with Theophrastus, not Alexander.[192]

Obertello 144 sees in Boethius' insistence that hypotheticals derive from categoricals a Peripatetic view, a hint that Boethius was using Peripatetic sources. But it could come from Galen, who (*Introduction to Logic* 4 and 14) criticized the third indemonstrable, or a contemporary of his. In addition, Apuleius (*DI* 192–3 Thomas) says that Aristotle and the Peripatetics used letters, but the Stoics numbers i.e. *primum, secundum*, etc. Boethius usually has letters, though he has numbers e.g. at 1 4 32–40. Speca (68–9) surveys these arguments with some scepticism and adds further references.

John of Salisbury, *Metalogicon* 4.4 p. 16.13–15 Webb is a mere summary of Boethius.

Abaelardus, *Dialectica introductio* p. 145 De Rijk 25–9 paraphrases **111A** 6–10.

1B Philoponus, *On Aristotle's* Prior Analytics 1.23 40b17 (CAG vol. 13.2 pp. 242.14–243.1 Wallies)

Speca 35–6, 38–42, 64–5,

The page number, 245, in the first impression was wrong. It has been corrected in the text title but not in the translation title in the second impression.

---

[190] The Kneales (180–1) postulate a Stoic philosopher who produced a set of seven rules similar to the rules of Chrysippus but also significantly different. This suggestion has given rise to a vigorous discussion, for which see Speca 107–11.

[191] This passage is studied by Speca 60–3.

[192] Ivo Thomas, (*Mind* 1955 274), reinforces this by comparing the language used by Boethius with that of Theophrastus. At **113B** 5 Alexander gives as one of Theophrastus' syllogisms "If man is, animal is, if animal is, substance is, if then man is, substance is." Boethius has: *est A, est B* etc., i.e. like Theophrastus putting in the verb "to be". Thomas sums up: "When substitutions are made in the two sets of propositional schemes, and the context is filled out in those of Theophrastus, identical results are obtained, which suggests something more Aristotelian than Stoic."

Most of the material here has already been covered in the introduction to hypotheticals.

ek deuterou is difficult: Repici fr.33b p. 140 has "*viene assunta per sostituzione per la seconda volta, visto che era stata gia posta nell' hegoumenon*". But it is not "*assunta per sostituzione*" for the second time. "As a second step" makes good sense, but is it possible with the Greek? Or one might write *ek prõtou* for *ek deuterou*, but in line 19 we have the definite article for both cases.

In Eudemus von Rhodos (Wehrli vol. 8) 53 it is suggested that Philoponus was following Antiochus of Ascalon here, and that accounts for the differences between him and others.[193]

pseudo-Themistius, *On Aristotle's* Prior Analytics 91.21–38
See after **108B**. This repeats Philoponus with few differences.

111C Farabi, *On Aristotle's* De interpretatione 4 17a6–7 (53.5–12 Kutsch and Marrow)

Zimmermann (1981) 45. Barnes (1985).

The translation of the last sentence of the fragment should be revised to read (after "books on logic"): "such [a separate treatment of hypothetical syllogisms] is found in the commentaries of the commentators only who give an account of them [hypothetical syllogisms] on the authority of Theophrastus."

Al-Fārābī's terse style in this passage has occasioned some discussion in secondary literature so it is important to be clear about what he is saying. The information al-Fārābī and his (anonymous) Greek source(s) are conveying is the following: (a) Aristotle did not write a separate treatise on hypothetical syllogisms; he just treated them slightly in *Prior Analytics* and not at all in *De interpretatione*; (b) Theophrastus and Eudemus, after Aristotle, treated hypothetical syllogisms exhaustively; (c) the Stoics, especially Chrysippus, treated hypothetical syllogisms exhaustively; (d) the Stoics[194] claimed that Aristotle wrote books about hypothetical syl-

---

[193] At 245.20 Philoponus has the example: The thing approaching is not both a man and a horse; but it is a man: therefore it is not a horse. But this, as not both p and q, but p, therefore not-q, is the third indemonstrable of Chrysippus, which is not supposed to have been used by the Peripatetics. That might be a sign that his source was eclectic.

[194] The subject of the plural verb *zaʿamū* can only be "the Stoics", *aṣḥāb ar-riwāq*; if it

logisms; (e) the separate treatment[195] of hypothetical syllogisms is to be found in the commentators who follow the treatment of Theophrastus.

As Barnes 140 notes, point (d) above "is the singularity" in al-Fārābī's account. This singularity can be explained only by assuming that al-Fārābī did indeed have access to 'some Greek source which we no longer possess';" Barnes' and Zimmermann's doubts on p. 140 are unfounded. For it stands to reason historically that Stoics could have made such a claim. Theophrastus and Eudemus personally knew about Aristotle's books and they could not have made any statements about a treatise on hypothetical syllogistic which Aristotle never wrote. However, after Aristotle's death his books were carted off to Asia Minor by Theophrastus, and though it now appears that significant knowledge of them is still in evidence in Attica, not all of them were readily available. Subsequent generations of Stoics, seeing that Theophrastus had written about hypothetical syllogisms, knowing that Theophrastus wrote on all the subjects that Aristotle had written on, and having no access to Aristotle's books for verification, simply assumed that Aristotle must have written about hypothetical syllogisms as well. This does not mean that the Stoics' claim was valid, merely that they thought so. The "singularity" in al-Fārābī's account thus provides us with a detail in the history of Greek philosophy which we did not have before. (DG)

D Galen, *Introduction to Logic* 3.3–5 (*BT* pp. 8.7–9.16 Kalbfleisch) trans. Kieffer 33–4.

Kieffer 66–70. S. Bobzien, 'The Stoics on Hypotheses' *Phronesis* 42 (1997) 299–312. Slomkowski 107–8. Speca 44–52. Fortenbaugh 2002 76.

---

had been either Theophrastus or Eudemus or both (as Barnes 140 and Zimmermann suggest), it would have been in the singular or dual. For the same reason, the plural verb *zaʿamū* cannot include the Stoics *and* Theophrastus and Eudemus, since neither of the latter two would have made such a statement. (DG)

[195] The Arabic has the demonstrative pronoun *ḏālika* ("that" or "such as what has been just mentioned"), referring to what immediately preceded, which theoretically can be either "the claim that Aristotle wrote a separate treatment of hypothetical syllogisms", or "a separate treatment of hypothetical syllogisms" (see the analysis of this sentence in Barnes (1985) 140). In the translation of **111C** I had preferred the former alternative ("this [claim]"); it now seems clear to me that the second is correct because, first, it fits the context better: in the understanding of the revised translation, the main and the relative clause talk about the same thing, the treatment of the hypothetical syllogisms, as opposed to the previous understanding. Second, the commentators, if they follow Theophrastus in the account of the hypothetical syllogisms, they cannot claim that Aristotle wrote an independent treatise on this subject because Theophrastus could not have said this when he dealt with it; as we saw, it was the Stoics that made this claim. (DG)

We have included this because although Galen does not mention Theophrastus he is clearly discussing the same subject-matter as the others we have quoted. His favoured term for the Peripatetics in this work is *hoi palaioi*:[196] he does not use *hoi archaioi*. He is here distinguishing the *palaioi* and the *neōteroi*, who must be the Stoics, in terms of the language they use about hypothetical syllogisms. The first type is "by connection"; the manuscript actually has *kata sunêthei(an)*, but Menas emended that to *kata sunecheian*,[197] following line 15–16 below.

The text of Galen is poor, and there are some crucial problems. The most serious of these is in line 12, where the sole manuscript has a lacuna, caused by the erasure of three letters after *ei* ("if"). The context implies that there was another conjunction beside *ei* and calls for a word of two syllables with the same meaning as *ei*, and we have adopted *eiper* (also "if") after Kieffer rather than *epei* ("since") from Prantl. We should also refer to **112C**, where Simplicius says that *epei* is used instead of *ei* when the antecedent is self-evident and undisputed.[198]

Other problems: In the first sentence Kieffer keeps *ê* ("or") and reads: "(or, as we said, 'by connection'). The statement is called hypothetical by the ancient philosophers". But since "hypothetical by connection" occurs below (lines 15–16), that seems unnecessary.

Barnes finds our rendering of lines 7–8 inadequate and suggests "just as the other kind of proposition, which we said comes about by connection, is called by them 'conditional'".

In line 23 we have printed *lexeōs* ("speech"), with Menas where *deixeōs* is a possible alternative, meaning a form of proof. But "If it is not night it is day" is not by itself a proof.

Speca has a thorough treatment of Galen and regards him as careful.

Galen, 14.2 (*BT* p. 32.11–15) trans. Kieffer 46.

Galen here refers to hypothetical syllogisms in a general way.

Ibid. 5.5 (p. 13.10–11) trans. Kieffer 36.

This is a straightforward study of the hypothetical by connection.

---

[196] "The ancients" is a fairly loose term, but its meaning is often obtainable from its context. David Sedley (1999) 140–3 considers its use by Proclus in *On Plato's* Parmenides about differing views about the proems to Plato's dialogues, and also discusses other uses. See also Barnes (1999) 36–8.

[197] Bobzien (2002) 59 points out that this terminology is rare, but finds it earlier in this chapter at 3.1.

[198] DL 7 71 reports that Crinis in his *Art of Dialectic* treated of the *parasunêmmenon*, a type of hypothetical proposition in which the conjunction used was *epei* ("since") which indicated that the antecedent was true. Crinis is of the late second century BC. See also **112C**.

E Alexander of Aphrodisias, *On Aristotle's* Prior Analytics 1.44 50a39 (*CAG* vol. 2.1 pp. 389.31–390.9 Wallies) trans. AlexPA (Mueller)[4] 84–5

Speca 1–3, 36–60.

Alexander here denies that there was in existence at his time any treatment of hypothetical syllogisms as promised in this passage of Aristotle.[199] Here and in **112BC** and **113C** we are told that it was in the first book of his *Prior Analytics* that Theophrastus treated of these matters. It is likely that that was available to our authorities.

Alexander ends with a list of what Aristotle's hypothetical syllogisms might be, after referring to Theophrastus, Eudemus, and others. His way of expressing himself is tentative, though the *legoi* we have rendered "he would be referring to" is an emendation by Wallies for the *legei* "he is referring to" of the MSS.[200] Even so, it is legitimate to ask what is the source of his list; his uncertainty suggests that he has not himself trawled through all the relevant works, nor does it look as if it has been taken from a handbook. Further, we cannot assign any of these types to particular authors. Fortenbaugh suggests that the list comes from Theophrastus and Eudemus, but it has Stoic elements. Slomkowski 104 takes lines 8–9 to imply that possibly Alexander had failed to find any of these syllogisms in Theophrastus. Speca 36–41 is doubtful about the view that Alexander is referring to Theophrastus, but that remains a strong possibility. We have taken the vague statement at the end "about which I have spoken elsewhere",[201] to refer to 324–5 above, for which see the apparatus.

Line 5: Barnes (1985) 126 suggests that the *tines alloi* might be Phainias or Strato.

Lines 8–9: this punctuation, proposed by Sedley, is confirmed, as Barnes notes, by **112A**.66–7. See also Barnes (1985) n. 2.

Lines 9–10: Fortenbaugh points out that here Alexander uses the expression *ex analogias* ("from analogy") which is different from the *kata analogian* ("by analogy") used of totally hypothetical syllogisms in **113A**. He suggests with some hesitation that it may have a different meaning, perhaps that of *PA* 1.46 51b5–28, commented on by Alexander at

---

199 Barnes suggests in line 2 "many other types of syllogism" for "many other conclusions".

200 On p. 38 Speca seems unaware that *legoi* is not the manuscript reading.

201 Literally "it has been stated", and so even vaguer.

397.25–398.15. Here Aristotle uses the term *analogon* of the relationship between two pairs, and Alexander uses the expression *di' analogias*. There is however no indication that this is anything to do with Theophrastus. See also Barnes 133–4.

For "qualitative" arguments see Barnes 132–3.

pseudo-Themistius on Aristotle's PA 152.15–23.

See after **108B**. Here a passage from Alexander 389.29–390.19 is placed between passages from Philoponus.

AlexPA 324.16–325.24

Alexander in **111E** refers back to this passage, in which he discusses at some length various kinds of arguments with a wealth of examples. He is commenting on *PA* 1.29 45b15, where Aristotle refers briefly to syllogisms *ex hupotheseôs*, like those *kata metalêpsin* and those *kata poiotêta*. For all this see on 113A–D below.

112–C Fortenbaugh, W.W., 'Cicero *On Invention* 1.51–77 Hypothetical Syllogisms and the Early Peripatetics', *Rhetorica* 16 1998 25–46 and in Fortenbaugh 2003.

These are on a form of argument in which there is one conditional and one assertoric premise, like "If P, then Q; but P; therefore Q." These are the forms on which the Stoics concentrated, and **112A** is largely concerned with comparing the Stoics and the Peripatetics, the younger and the older men respectively. It is generally agreed that the earliest Stoic logician of any standing was Chrysippus, so that whatever work on the subject that Theophrastus did must have been original.

112A Alexander of Aphrodisias, *On Aristotle's* Prior Analytics 1.23 41a37 (*CAG* vol. 2.1 pp. 262.28–264.31 Wallies) trans. AlexPA (Mueller)[3] 40–43.

Speca 47–8.

Theophrastus is mentioned only once, in line 24, and the information about the Stoics can hardly be from him. Even so, we have chosen to print much more, to cover both Stoic and Peripatetic accounts. The source is uncertain, and it may reflect Alexander's concern with the categorical syllogism. But we may take it that in his own *Prior Analytics*, in the first book, Theophrastus devoted considerable attention to a subject that Aristotle had only touched on.

Barnes has queried the sense of *hôs eipen* in line 1, which should mean "as he said" and we have taken as meaning "his words"; Aristotle did not precisely use the words Alexander gives, but at 1.23 41a40–1 he used a very similar turn of phrase, "by agreement or some other hypothesis". The derived text in pseudo-Themistius has *hôs eipon*—"as I said", which is out of place in that work and difficult to explain, unless the author was copying an original *hôs eipon* in Alexander. But even that could have come from a faulty text.

Line 5: Bobzien 2002a 381 n. 55 suggests that *deiktikê*, for which she uses "probative", was the term used by Theophrastus for what by Alexander's time was called "categorical". See also n. 81.

Line 19: Barnes suggests that instead of "scope" for *chreia* it would be better to put "function" or "utility".

Lines 23–5: Fortenbaugh relates this and **112B** to the fivefold analysis of deductive reasoning attributed to Aristotle and Theophrastus by Cicero *On Invention* 1.51–**77**. But he points out that in this early work Cicero is unlikely to have known Theophrastus directly, and probably got his material from rhetoricians who themselves used the name of Theophrastus merely as a known follower of Aristotle.

Kieffer 129 suggests that in line 46 Alexander is taking *metalêpsis* as a minor premise of a hypothetical, and *proslêpsis* as that of a categorical syllogism, and at 67 that he uses terminology that originally had another reference to talk in Peripatetic terms.

Line 48: for "in a relative form and" read "in the relationship of".

Line 57–8: There is something wrong with the text here. We need either an expansion, for which Barnes suggests inserting ἀλλὰ μὴν οὐ τόδε· τόδε ἄρα ("but not this: this then") in line 57 after the second τόδε, thus giving a complete argument-form, or a deletion of ἤτοι· τόδε ἢ τόδε ("either this or this") leaving the bare description. The word *tropikos* is unusual, but its meaning is made clear by Alexander. Galen (7.1) says that that word is used because the whole syllogism "is pegged upon them as the timbers upon the keel of a ship." (Kieffer's translation). Kieffer considers this etymology on p. 93. All we can be sure of is that the word is rare and its derivation uncertain. See now also AlexPA (Mueller)[4] p. 26 n. 11. In line 67 Alexander uses *tropos* of the Stoic moods. For a general discussion of Aristotle's proof see Alexander on the *PA* 256 9–18 and 386.5–387.27.

Parts of this passage[202] are in Graeser's Frs. 10 (on prosleptics), 28 and 29.

[202] Barnes has noted the following corrections: line 33 κἀκεῖνο, 36 πᾶσα, 63 οἷον.

pseudo-Themistius *On Aristotle's* PA. 73.19–75.15.

See on **108B**. This is part of a passage based on Alexander 255.3–267.4 with minor differences.

Scholion on Aristotle's PA (*CAG* vol.4.6 xi.30–2):

This is in a long account of hypothetical syllogisms. Here we have the negative conjunction with the example: 'He is not both at Athens and at Megara: he is at Athens, therefore he is not at Megara'.

Philoponus *On Aristotle's* PA p. 245.23–4

Philoponus says that the third *tropos* is that from a negative conjunction, affirming the one and denying the rest.

112B Alexander of Aphrodisias, *On Aristotle's* Prior Analytics 1.44 50a16 (*CAG* vol. 2.1 p. 388.17–20 Wallies) trans. AlexPA (Mueller)[4] 93

In the footnote, for "conclusion" read "premise".

This occurs in a passage where Alexander is writing about Aristotle's denial that one can reduce hypothetical syllogisms to tne figures. Before bringing in Theophrastus he refers to various cases: 1) an argument *ex hupotheseôs* ("from a hypothesis") (388.7), 2) self-evidence (388.7–8 and 14–17), 3) by induction (388.8–9). The cases are all connected with Aristotle's discussion of health and sickness, so that it is unlikely that Alexander has taken them over from Theophrastus, but the possibility is there.

In any case we have solid evidence about Theophrastus' treatment of this matter. In **112A** the only direct information was that he said that the additional premise was doubtful and needing to be proved. Here he is said to have gone into detail about the various ways in which the additional assumption or premise may occur. This appears to be in conflict with Alexander's account in **112A**, which appears to claim that the additional premise can be established only by means of a syllogism, of which Alexander gives several examples. Self-evidence is associated with Theophrastus by Sextus Empiricus in the account of Peripatetic epistemology which we give at **301A** (*Against the Mathematicians* 7 216–26). At our line 6 (**217**) he says that Theophrastus said that self-evidence was common to both sense and intellectual activity (*noesis*). The idea of self-evidence occurs also in **112C**, there applied to the antecedent, but Theophrastus' position here is not clear.

On our evidence we may see Theophrastus as turning to the minor premise or additional assumption of a mixed hypothetical syllogism and saying that it is doubtful and needing proof. But that cannot be the whole story, for here in **112B** he is giving a range of possibilities. That is what one would expect. If he did take hypotheticals seriously as important for reasoning he would have seen the variety of types involved.

2C Simplicius, *On Aristotle's* On Heaven 3.1 298a27 (*CAG* vol. 7 pp. 552.31–53.4 Heiberg)

Speca 56–7.

For Simplicius'career see Blumenthal (1996) 42–6.

For *ei* and *epei* see on **111D**. The explanation of the difference is clear, but what Theophrastus said is not. The terms parasynaptic and synaptic are technical and not likely to have been used by the earlier Peripatetics, so that he can hardly have used them.[203] On the other hand it is difficult to work out what he could have said except what Simplicius reports, which seems inadequate. What explanation do we need for the fact that two different words fulfil two different purposes?

Fortenbaugh 38 (and 64) suggests that Theophrastus defended the use of *epei* for *ei* on the grounds of economy, for *epei* both indicates a latent premise and also a categorical statement. He also suggests (n. 48) that Theophrastus anticipated the Stoics with their parasynaptic (DL 7.71) in which the protasis is a fact.

Boethius, *On Hypothetical Syllogisms* 1.2.5 attributes to Eudemus alone the view that *cum homo est animal est* has the same force (*eaedem potestatis*) as *si homo est animal est.* This may indicate that Eudemus and Theophrastus differed on this point,[204] but Simplicius is vague about what Theophrastus actually said.[205]

In the margin write 553 for 533.

Sextus Empiricus, *Against the Mathematicians* 8 109–10

Sextus is speaking of the conditional (*sunêmmenon*) proposition form with *ei* or *eiper*, with *ei* used in a proposition involving repetition, like "If it is day it is day", and *eiper* when the elements differ, as with "If it is day, it is light".

---

[203] DL 7 71 reports that Crinis, in his *Art of Dialectic* treated of the *parasunêmmenon,* a type of hypothetical proposition in which the conjunction used was *epei* ("since"), which indicated that the first proposition was true. Crinis is of the late second century BC, and seems to have been a maker of definitions.

[204] Bobzien (2002a) 66–8 considers at length Boethius' evidence about Eudemus, and at 70–1 studies Boethius' whole account of hypothetical syllogisms to point out its similarity to that of the early Peripatetics. At 2002b 377–80 she also studies Eudemus' account of the meaning of *hypothesis* given by Boethius.

[205] See the reading of Prantl/Kalbfleisch at Galen *Introduction to Logic.* 8.20 which supplies *ê epei* after *ei.*

113A–D In *PA* 1.29 45b15–20 Aristotle refers to syllogisms *ex hupotheseôs*, ("from a hypothesis") like those *kata metalêpsin* ("by substitution") or *kata poiotêta*, ("by a qualitative relation"),[206] and says it is necessary to decide how many kinds of such syllogisms there are. At 1.23 40b23–41b6 he discusses kinds of proof, and refers to proofs *ex hupotheseôs*. It is clear that he expects his readers to understand that expression. Finally there is a promise to deal with syllogisms *ex hupotheseôs* systematically at 1.44 50a39–b1, a passage which raises certain questions. Our evidence tells us that Theophrastus developed the theory of totally hypothetical syllogisms and related them to the figures of categorical syllogisms, in which unprofitable exercise he was followed, with a difference, by later logicians.[207]

113A Alexander of Aphrodisias, *On* Aristotle's Prior Analytics 1.29 45b19 (*CAG* vol. 2.1 pp. 326.8–12 Wallies) trans. AlexPA (Mueller)[3] 116–7.

It has been asked whether we should have extended this passage to line 19, thus filling in the gap between **113A** and **113B**, but those lines seem part of Alexander's general argument about hypothetical syllogisms.

In chapter 29 Aristotle has been dealing with methods of proof by reduction and by ostensive proof. He then passes to the cases he describes as "by hypothesis" names some and says that it is necessary to have a division to see how many of these there are. Alexander brings in the wholly hypothetical syllogisms, perhaps because these only do not prove the existence of something. At the end of **113D** it is stated that Alexander denied that wholly hypothetical syllogisms were syllogisms at all because they displayed only consequence.

For the terms used see on **113C**.

113B Alexander of Aphrodisias, *On Aristotle's* Prior Analytics 1.29 45b19 (*CAG* vol. 2.1 p. 326.20–31 and 327.2–5 and 15–16 and 33–328.6 Wallies) trans. AlexPA (Mueller)[3] 117–9.

Speca 29–32.

---

[206] Maier 2a 282–4 suggests that all this is an interpolation 3using terms from Theophrastus, but that is unnecessary. See Ross 395.

[207] Barnes (2002) 121–3 discusses in Theophrastus in connection with a Byzantine logical document.

Alexander first uses letters, A, B, and C, to make his points, but then gives an example with man, animal, and substance. In the passages omitted from our quotations he gives five forms, and a sixth by implication, which go into three groups:

1a) A implies B, B implies C therefore A implies C.
1b) A implies B, B implies C, therefore not-C implies not-A.
2a) A implies C, B implies not-C, therefore A implies not-B
2b) (and by implication) B implies not-A
3a) A implies B, not-A implies C, therefore not-B implies C
3b) A implies B, not-A implies C, therefore not-C implies B

Both Theophrastus and Alexander related 1) to the first figure but Alexander gives long arguments for relating 2) to the second and 3) to the third.

See Bobzien (2000) for why the order of the figures was changed. Alexander gives a clear account of the reasons for his view.

The meaning of lines 22–4 is not clear. After them Alexander immediately turns to the views of Theophrastus about the order of the syllogisms, introducing them with *mentoi* ("however") as if they were related to 22–4; it seems better however to take 22–4 as an independent summary and conclusion to what goes before, with the remarks about Theophrastus intrrpolated awkwardly. Alexander then says that it would be opportune to write separately about these matters, but we do not know if he ever did.

Alcinous, *Didascalicus* 6 (p. 159.10–12 Hermann)

For a long time there was doubt about whether Albinus or Alcinous was the author of the *Didascalicus.* Recent work has assigned it firmly to Alcinous. He is here finding hypothetical syllogisms in Plato. He says that there are three figures and admits that people differ over the second and third figures. See Alcinous, *The Handbook of Platonism* translated with Introduction and Commentary by John Dillon, Oxford: Clarendon Pess 1993 81–3.

Boethius *On Hypothetical Syllogisms* 2.9.1–3.6.4

Boethius discusses "intermediate connective statements".

C Philoponus, *On Aristotle's* Prior Analytics 1.29 45b18 (CAG vol. 13.2 p. 302.6–23 Wallies)

For Philoponus see on **71E**.

Philoponus is reporting Alexander's views, dealing with his 326.8–10, where he says that the totally hypotheticals do not fall under proof by selection of terms (**113A**). In line 8 Philoponus uses the word *akolouthia*, which we have rendered "consequence". But Alexander does not have that word in this context. See Hintikka (1973) 43–7 for a survey of its uses, in which he finds at least three senses in Aristotle.[208] Kieffer 76–82 discusses the use of *akolouthia* in Galen. In general the sense has to be worked out from the context.

Dürr 7 doubts if Philoponus knew directly the sources that Alexander used. Philoponus, below 243.14–17, says that these syllogisms are called *dia triôn* and *di' holôn hypothetikoi*, the latter because all the propositions are hypothetical, the former because they are completed through at least three "hypotheses" ("conditionals"). He gives no names here, and his examples are about God, the afterlife, and other eschatological topics.

Dürr doubts if the term *di' holou hupothetikon* is correct, and prefers Alexander's "by analogy". It cannot be decided, I think, whether Theophrastus used *di' holou hupothetikon*, which is not found in Aristotle: Alexander (**113AB**) uses that expression freely, as if this were the accepted term for these cases. Yet it is not he, but Philoponus, who credits Theophrastus with it. Philoponus is clearly using Alexander, and it seems likely that he is misusing him here. The special term invented by Theophrastus was "by analogy". Mueller in AlexPA (Mueller)[3] p. 151 n. 253 suggests that there may be two analogies involved, that between the items of the syllogisms themselves, and that between these syllogisms and the ordinary categorical syllogisms, relying on **113C** 12–14. Alexander also uses "*dia triôn*", without indicating its source.

AlexPA 326.12–19

This passage precedes **113B** in Alexander. Alexander sets out the view that totally hypothetical "syllogisms" are not syllogisms at all, because they do not prove the existence of anything.

113D Scholium on Aristotle's *Prior Analytics* 1.29 45b18 (cod. Flor. bibl. nat. cent., Conv. Soppr. J 6.34 = *Arist. Lat.* vol. 3.4 p. 320.7–16 Minio-Paluello)

This is in a beautiful copy of Boethius' second translation of the *PA* with many scholia in the same hand. Minio thinks they are Boethius' own scholia. Minio 'A Latin Commentary (translated by Boethius) on the *Prior Analytics* and its Greek Sources'. Minio (*Aristoteles Latinus* vol. 3 pt. 4

[208] See also 188–9 for Diodorus' usages.

Bruges/Paris: Desclée de Brouwer 1962) xxxvii thinks that the similarity with Philoponus is due to their drawing on the same sources.

The statement of Aristotle's against which Theophrastus was arguing must be that the same method (*tropos*) of arguing applies to hypothetical as to categorical syllogisms. It is not clear whether it is the scholiast who relates the two directly, or whether Theophrastus himself did this in his *Prior Analytics.*

The word *consequentiam* here must be the equivalent of the Greek *akolouthia* for which see on **113C**.

There is a passage on hypothetical syllogisms reported by Waitz in his Organon I 9–10, little known because it is printed in the descriptions of the MSS and not in the scholia. It is at the end of the *PostA* in the important eleventh century MS Laurentianus 72,5 (Ross's d). It uses Stoic examples, but allocates the syllogisms to the three figures and in Theophrastus' way.

It sets out briskly three kinds of "hypothetical" arguments, those beginning with a *sunêmmenon* (conditional), those with a *dialelumenon* (disjunctive), and those which we have considered under the title "wholly hypothetical". All are set out in detail, and the wholly hypothetical are assigned to figures in Theophrastus' way. See *LCM* 4 (1979) 207–9 for a translation.

It may well be, as A.C. Lloyd thought (private communication), a Byzantine work, but its vocabulary is unusual and could come from a time before a standard vocabulary had been established:

1) *hupothetikoi sullogismoi* ("hypothetical syllogisms") is the standard expression after Aristotle, who used *ex hupotheseos* ("from a hypothesis").

2) οἱ ἐκ δύο ὅρων συνημμένων ἢ διαλελυμένων ("those from two terms joined or divided"). This way of expressing the nature of conditional and disjunctive propositions is unusual. Throughout the work the notion of *horos* as part of a proposition is prominent. It is *horoi* that are joined or divided, whereas elsewhere it is the whole proposition that is called "joined" etc.[209] At *PA* 1.1 24b16 Aristotle himself seems to allow for this usage in defining *horos* as "that into which the protasis is divided (*dialuetai*)". Further, he sometimes uses A and B in this way.

3) *dialelumenon* This word, which clearly refers to what the Stoics called *diezeugmenon* does not seem to survive. Martha Kneale suggested it was a

[209] At line 25 there is the standard form.

manuscript error, but as it occurs twice that seems unlikely. The Peripatetic term reported by Galen and Alexander is *diairetikon*, but that is not directly attributed to Theophrastus.

4) *epanodos* This term is used by Aristotle in his *Rhetoric* 3.13 1414b2 to mean recapitulation (of an argument). It is appropriate here, as the repetition of a set of words, but the use is not found elsewhere. Waitz says it is equivalent to *metalêpsis* in Aristotle.

5) *kata thesin* and *kata anairesin* ("by affirmation" and "by denial") These expressions are not in Aristotle, but he did use e.g. 2.14 62b30 *tithenai* ("to affirm") and *anairein* ("to deny").

6) Aristotle did not use *ho sunagôn horos* ("the linking term") instead, as Waitz says, he spoke of the *mesos* ("middle") term; *ho agôn horos* is not found elsewhere; nor is *hoi sunagomenoi* two lines above.

7) *lêgôn* This standard Stoic term for "consequent" seems here to refer only to the position of the term. The word lêgein means to cease, particularly to cease speaking, and it is therefore a suitable expression for the second part of a conditional. It seems to have only this limited sense here. So it could be that Theophrastus used it in this way first, and it later became attached in particular to the consequent, and so came to mean it.

8) *anapodeiktos* ("unproved") and *analusis* are both connected with Aristotle's theory of first figure syllogisms to which the others are reduced.

9) Martha Kneale has objected to the use of *to A, to B* ("the A, the B") etc. here, on the grounds that the Peripatetics used letters and the Stoics numbers, and that this is a hybrid in which the A and B are numerals. But Alexander in **113B** line 4 has exactly this usage in his account of Theophrastus' hypotheticals.

I conclude that this is a curious work which may contain traces of early ways of speaking.

A possible gap in our knowledge of Peripatetic thought at this time is indicated by a passage in Sextus Empiricus, *Outlines of Pyrrhonism* 2 110–12 which tells of a controversy between Diodorus Cronos and his pupil Philo,[210] who were both contemporaries of Theophrastus, about the

[210] It is not unreasonable to identify this Philo with Philo of Athens, a disciple of Pyrrho, who is known to have written a *Menexenus* (Clement of Alexandria, Strom.IV 528). Aristotle as well as Plato wrote a *Menexenus* and that suggests a connexion with the Academy.

analysis of the conditional "If p then q". Besides these Sextus refers to two other theories, one of which has been attributed to Chrysippus.[211] The fourth has been ascribed by Martha Kneale (1962) 129 to a Peripatetic source. The view, ascribed to "those who judge by *emphasis*",[212] is that the consequent must be potentially in the antecedent. With this, it is said, a form like "If it is day, it is day" would be excluded. Unfortunately no acceptable example is given.

The order of the theories given appears to be logical rather than chronological, in that the later one always rejects the proposition accepted by its predecessors. All we can say is that the first two views were being canvassed in the lifetime of Theophrastus, and probably in Athens. There were not many Peripatetic logicians of the relevant period, so that Theophrastus is at least a candidate.[213]

At this point we must consider the relationship between Theophrastus and Stoic logic. It has traditionally been supposed that they were unrelated, but there are considerable similarities in the area of hypotheticals, and there was close proximity in space and time between their schools. Zeno and Theophrastus were both living and teaching in Athens from 312 when Zeno arrived there to 286 when Theophrastus died. There are highly coloured accounts of rivalry between the various schools, but it is improbable that men seeking the truth would limit their intercourse to polemical argument alone. And even if the masters were hostile, the students might pass from one school to another. There is indeed a chain linking Chrysippus, the great master of Stoic logic, with Theophrastus. Arcesilaus[214] studied under Theophrastus before joining

---

[211] Josiah B. Gould, 'Chrysippus on the criteria for the truth of a conditional proposition', *Phronesis* 12 (1967) 152–61 points out that it is only by lengthy argument that it can be shown to be by him. In his *The Philosophy of Chrysippus* he devotes six pages to an attempt to show that it is in fact Chrysippus' view. In the course of the argument he makes a number of assumptions, and it seems to me open to doubt whether the correct attribution has yet been discovered. M. Frede, Die *Stoische Logik*, G ttingen: Vandenhoeck und Ruprecht 1974 90–94 covers suggestions in great detail.

[212] Gould (1967) 153 renders this "implication'. M. Frede 90–1 says that this fourth definition is unknown elsewhere, and argues with examples that the attribution to the Peripatetics is improbable.

[213] To pile speculation on speculation, I suggest a possible home for all this in the *Megarikos* (**1** 129), discussed at the end of **68**. All we know of that work is that it tells of Diogenes of Sinope taking lessons from the behaviour of a mouse. Diogenes died in 323, and spent most of his time in Athens and then Corinth. But he did visit Megara (DL 5 410). At DL 6 69 he is shown using an argument of the form If P then Q, but P; therefore Q.

[214] See J. Glucker, 'Theophrastus, the Academy, and the Athenian Philosophical Atmosphere', in Van Ophuisen and van Raalte 299–316 for the background here.

the Academy, and Chrysippus studied under Arcesilaus before becoming a Stoic. Since the Academy appears to have accepted Peripatetic logic, Chrysippus may have studied it in his youth.

Barnes, 'Aristotle and Stoic Logic' in Ierodiakonou (1999) 37–8 argues that Chrysippus had read Theophrastus' work *On the Liar*, and (45–9) points out that he has the same logical particles, not, or, if, and and, as Theophrastus, but he rejects the view that Chrysippus used Theophrastus' logical system.

The following sections, on Demonstration and Definition, are related to Aristotle's *PostA*. We have very little from Theophrastus' *Posterior Analytics*, perhaps because we do not have Alexander's *Commentary* on Aristotle's work of that name.[215] But since Theophrastus' work bears the same title, it is reasonable to suppose that here too he followed Aristotle.

Demonstration

114 Galen, *On the Doctrines of Hippocrates and Plato* 2.2.3–5 (*CMG* vol. 4.5.1.2 pp. 102.25–104.8 De Lacey)

Galen is engaged in one of his tirades against other physicians, and this tells us little beyond the fact that Theophrastus as well as Aristotle wrote a work called *Posterior Analytics.* As we have said, that of Theophrastus must have received its title after that of Aristotle had been named, and although it might have been written well before it received that title it must have been based on the latter. Galen treats them as doctrinally very similar.

line 5: ὑπὲρ αὐτῆς ("on it") is ambiguous, but surely must refer back to τὴν ἀποδεικτικὴν μέθοδον ("the method of demonstration") of line 3.

115 Themistius, *Paraphrase of Aristotle's* Posterior Analytics on 1.2 72a8–24 *CAG* vol. 5.1 p. 7.3–6 Wallies)

The relevant passage of Aristotle covers a number of matters. It is not obvious why Themistius should have chosen to refer to Theophrastus at this point, but it could be that it was here that Theophrastus decided to address these matters, bringing in material from other parts of Aristotle.

[215] For its fragments see Moraux 1979.

The word *axioma* here is, as one would expect, used in the Peripatetic, not the Stoic, sense. For "if equals from equals" see Aristotle *PostA* 1.10 76a41 etc and Euclid *Elements* 1.3. The whole axiom is: "if equals (are taken) from equals, the remainders are equal". But for "the affirmation or the negation" the position is less clear. It should be related to the basic rules of *Metaphysics* Gamma and mean that either the affirmation or the negation must be true, i.e. to the law of excluded middle; but it is not easy to find such a principle in this form in Aristotle. At *Metaphysics* K 1061b17 there is a reply to a question raised at the beginning of this book (1059a23–6), whether it is the business of a single form of knowledge to study the axioms. Here there is a contrast between the mathematician who uses "if equals are taken from equals, the remainders are equal" (quoted in full), and philosophy, which studies being *qua* being. This part of book K corresponds to Gamma 1005a19–b2, but the axiom about equals is not mentioned there. See *PostA* 1.11 77a30–31 which cites the two axioms in a very shortened form.

There is also a puzzle about why Themistius gives the opening words of the supposed axiom in the accusative form. That is not required by the context, and suggests, perhaps, that there was a list of axioms of which this one was expressed wih an accusative and an infinitive.

Cod. Vat. Gr. 246 f.12[v]

This part of the 13th to 14th century ms. contains definitions taken from Aristotle's logical works.

6 Philoponus, *On Aristotle's* Posterior Analytics 1.4 73a27 (*CAG* vol. 13 p. 71.4–14 Wallies)

Mignucci *L'Argomentazione demonstrativa in Aristotele* Padua: Antenore 1975 81–2. J.E. Tiles. 'Why the Triangle has Two Right Angles *Kath' Hauto*' *Phronesis* 28 (1983) 1–30.

For Philoponus see on **71E**.

Ross 523 comments on this part of Aristotle "[Aristotle] is making his terminology as he goes. Having first used *kath' hauto* and *hêi auto* as standing for different conditions, he now intimates that *kath' hauto* in a stricter sense means the same as *hêi auto*." The point here is that what is *hêi auto* is not only necessarily true of a subject but also exclusive to it: what is *kath' hauto* is also necessarily true, but may be shared with other subjects. What is *hêi auto* is therefore also *kath' hauto*, but not necessarily *vice versa*.

Barnes² 122 argues that Theophrastus differed from Aristotle over the interpretation of a statement like "Every isosceles has an angle-sum of 180°". For Aristotle this cannot be, in Barnes' terminology, an I-predication, i.e. one in which, for example, A holds of B "in itself". But Theophrastus made a distinction between "as such" and "in itself". "We must infer that Theophrastus gave a new sense to I-predications... it is reasonable to conjecture that it was designed to insure that *all* scientific propositions, theorems as well as principles, may be classified as I-predications."

117 Alexander of Aphrodisias, *Questions and Solutions* 1.26 (*Suppl. Arist.* vol. 2.2 p. 42.27–31 Bruns) trans. Sharples, Alexander of Aphrodisias, *Quaestiones* 1.1–2.15 London: Duckworth 1992 (*ACOA*) 89.

Sharples (above) 86–90.

In the second impression we have substituted for "that was said ... in the sense" "that, in the definition of which there is included something to which it belongs, was said to belong *in itself* to (that thing)—in the sense"....

This is from *Quaestio* 1.26, 'How form is in matter, whether in itself or accidentally". That is considered at length, but without final decision. So in our passage Alexander brings in the distinctions of various kinds of "in itself" discussed by Aristotle at *PostA* 1.4 73a34. The suggestion is that it is possible for a form to be in a subject "in itself" but not in its being. Alexander gives as examples the way odd and even number are in number. This is but one of the examples given by Aristotle, and the fact that Alexander speaks of Aristotle and Theophrastus together, with a single reference to the *PostA*, suggests that the difference between them was not great.[216]

Definition

118 Boethius *On Aristotle's* De interpretatione, 5 17a8 Second edition, 2.5 (*BT* pp. 102.26–103.11 Meiser)

[216] I have benefited from correspondence with both the late A.C. Lloyd and Bob Sharples about this passage.

A.C. Lloyd, *The Anatomy of Platonism* Oxford: Clarendon Press 1990 56–8 thinks both Boethius and Ammonius get this from Porphyry. Ammonius names Theophrastus several times in his commentary, though not here. See the reference in the Apparatus.

Graeser relates this to *DI* 5 17a13 and compares *PostA* 2 10 and *Metaphysics* Z.12 1038a31–3. In spite of Graeser and Repici (fr. 9), who give only the first four lines, we follow Wimmer in taking the quotation down to 103.11.

Boethius ascribes this point to Theophrastus' *On Affirmation* and there is no reason to doubt that. Aristotle had considered the unity of an utterance, e.g. statement or definition, in various places in his *DI*, in chapters 5, 8, and 11. Chapter 8 can be ignored, but in 5, though it is mainly concerned with what is a simple statement, he refers to definition at 17a3 "It belongs to another enquiry to say why footed two-footed animal (*zôon pezon dipoun*) is one thing and not many: at any rate it will not be one in virtue of being said all together", and in 11 he says that there will be a single affirmation or denial when one is said of many, or many of one, only if the many make up a one, e.g. man is perhaps animal and two-footed and tame, and these make up one thing, but white and man and walking do not make up one thing. Aristotle's own position is not entirely satisfactory, but he is touching on important matters. See Ackrill 145–9.

De Rijk (1962) 39 suggests that Theophrastus, on *DI* 17a35–7 referred to the definition of elenchus at *Soph. El.*5 167a23–7. He argues that Boethius had access to Theophrastus through Porphyry's commentary on Theophrastus (See on **80**).

He thinks it also possible that the six moods (*tropoi*) of fallacies given by Boethius also go back to Theophrastus. Unfortunately no early commentary on the *Sophistic Refutations* exists. Pseudo-Alexander is very late, but in some ways resembles Boethius in his examples of fallacies. Ammonius is different.

Boethius' first approach is metaphysical: there is one *oratio* because it points out one subject. See *Metaph.* Z 12 1037b24–7. *Metaph.* Z 12, H 6, and I 9 are relevant. But he then goes on to the unity of the utterance: if one breaks up a definition by means of intervals in speaking it becomes a multiple utterance.

In his more elementary discussion at *On Aristotle's* DI[1] 151,5 Meiser Boethius, commenting on Chapter 11, says: *multa Peripateticis de discernendis propositionibus quae essent unae, quae multae consideratio fuit,*[217] ("the Peripatetics gave much consideration to distinguishing propositions that were unitary and those that were multiple"), and what he goes on to say relates as much to the unity of a definition as to that of a statement. This implies that there was considerable interest in these matters. Theophrastus must at least have said that the unity of a definition is not the same as the unity of the utterance in which it is expressed, for that could be broken up by accident.

After our passage, Boethius says that it can be objected to Theophrastus' account: "if it is not uninterrupted utterance that makes one thing what is naturally manifold, why should an *oratio* which is one become manifold just by a delay in completing its utterance?" Boethius' commonsense reply here is that if the parts of a definition are uttered separately it is not as if they were referring to one subject.

It is impossible to say how much of this and what follows can be attributed to Theophrastus or one of his contemporaries.

Ammonius, *On Aristotle's* DI.5 17a19 (pp. 70.30–71.32) trans. AmmonBlank 77–8.

Ammonius, probably following Proclus, is commenting at length on *DI.* 17a9, and in particular deals with the unity of a definition.

Siger de Courtraco *On Aristotle's* DI.5 17a13 (p. 43.30–1 Verhaak)

For Siger see apparatus to **82B**. He followed Boethius.

William of Ockham *On Aristotle's* DI.1.4.3

William reported the same view without naming Theophrastus. Guillelmus is the spelling of the Franciscan Institute.

Boethius, *On Aristotle's* DI first edition 11 p. 151.20–2.

119 Eustratius, *On Aristotle's* Posterior Analytics 2.3 90b23 (*CAG* vol. 21.1 p. 44.1–7 Hayduck)

For Eustratius see R. Browning, 'An Unpublished Funeral Oration on Anna Comnena' and H.P.F. Mercken, 'The Greek Commentators on Aristotle's Ethics' both in Sorabji (1990) 399–401, 410–14. They differ over his dates. He was a pupil of John of Italy. See A.C. Lloyd, 'The Aristotelianism of Eustratios of Nicaea' 596 in Wiesner 1987 II 341–51.

---

[217] Boethius refers back to this at 352.2 on chapter 11, *nos autem supra iam diximus magnae fuisse curae apud Peripateticae sectae principes diiudicare quid esset adfirmatio vel negatio, quae plures.* ("We have already said above that the leaders of the Paripatetic school took great care to decide what affirmation was and what denial, and what plurality").

This commentary is only on book 2 of the *PostA.*

Eustratius has pages and pages on these bits of Aristotle. He could here be saying: "Perhaps (i.e. on a different view which I do not necessarily share)", and we then have an attempt to construct an argument which would make sense of what Aristotle says. So he starts with Aristotle's own words: "of one as one there is (only) one episteme." He then imports from Theophrastus and Alexander the statement that there are two modes of understanding, definition and proof. The conclusion is that there can be only definition or proof of one thing as one, not both. The point that there are two modes of understanding is not made in this part of the *PostA,* and we do not know in what connection Theophrastus may have said it. Aristotle frequently uses the expression *apodeiktike episteme* (understanding by proof) e.g. *PostA* 1 3 73a22, but never *horistike* (by definition).

Eustratius goes on to some more difficult arguing, but it does not proceed directly to a conclusion that to understand what is to be proved is to have the proof, and does not help us further about Theophrastus.

It is clear in Aristotle that he is not saying that these are the forms of knowledge in the strict sense, but rather that definition and proof are independent. It is not clear why Theophrastus and Alexander are brought in, but as in **120** Alexander is said to have cited Theophrastus, that could be the case here too. Eustratius may have thought that the chapter was a muddle.

It has been suggested that for *kuriôs* (properly speaking) in line 3 *kurias* (proper) should be read, but that seems unnecessary.

120 Leo Magentinus, *Introduction to Aristotle's* Posterior Analytics 2 (cod. Paris. Gr. 1917 f.257v and again 272v = p. 240a46–7 and b1–4 Brandis)

The work of Leo Magentinus was originally accepted for the Berlin Academy edition of commentators (CAG), but later rejected.[218] Magentinus was of the thirteenth century.[219]

Line 2 is ambiguous. It could mean that Alexander believed that definition was Aristotle's primary interest, or it could mean that he believed that it was not his primary interest. The former seems preferable. In

[218] See K. Praechter, 'Review of the Commentaria in Aristotelem Graeca' (A translation by Victor Caston of the original article in *Byz. Zeit.* 18 (1909) 516–36) in Sorabji (1990) 53.

[219] See Busse, *CAG* IV.5 xxxli for some account of Magentinus.

any case we can deduce that Theophrastus discussed this matter, and presumably gave reasons for the conclusion he arrived at.

The statement in the apparatus that MS C has *ou* after *zetei* is not in Brandis but is given by Moraux (1979)

121 Ptolemy, *The Will of Aristotle, the Pinax of His Works and a Brief Life, to Gallus*, from the Pinax (cod. Istanbul, Ayasofya 4833, f.16v 12–13.

Badawī (1968) 183. Düring, 'Ptolemy's Vita Aristotelis Rediscovered,' in R.B. Palmer and R. Hammerton-Kelly, eds, *Philomathes* [Festschrift P. Merlan], The Hague: M. Nijhoff, 1971, 264–269. M. Plezia, 'De Ptolemaeo pinacographo,' *Eos* 63 (1975) 37–42. Hein (1985) 388–439. D. Gutas, 'The Spurious and the Authentic in the Arabic Lives of Aristotle,' in J. Kraye, W.F. Ryan, and C.B. Schmitt, eds, *Pseudo-Aristotle in the Middle Ages: The Theology and Other Texts*, London: The Warburg Institute, 1986, 15–36; repr. in Gutas (2000) no. VI.

Ptolemy's *Pinax* survives only in Arabic translation in direct transmission in a unique MS (as above), and indirectly in the biographical works of Zawzanī (Qifṭī) and Ibn-Abī-Uṣaybiʿa. In many instances the text in the indirect transmitters is more accurate (i.e. Qifṭī and Ibn-Abī-Uṣaybiʿa had access to a better MS of the Arabic translation of Prolemy's *Pinax*) than that in the Istanbul MS that has preserved the directly transmitted text. In establishing the text of the title in this fragment, the readings in the MS have to be corrected according to the text in Ibn-Abī-Uṣaybiʿa, as I have done in my edition. Zawzanī (Qifṭī?) omits this title. A critical edition of Ptolemy's text is still a desideratum.[220] In the meantime, the references listed above ought to be consulted. (DG)

This claims to be the title of a work by Aristotle. Baumstark conjectures that the Greek was *technê horistikê*, but that title is not attested anywhere else for Aristotle. In any case it is difficult to relate this tale to what we know elsewhere about Theophrastus' *Prior Analytics.* See n. 73. (PMH)

---

[220] See my remarks in the article cited above, p. 29. See also notes 30, 31, and 58 in the same article for further information about the manuscript and the state of secondary literature on Ptolemy's *Pinax*. (DG)

Topics

Theophrastus accepted Aristotle's approach to topics, for Averroes tells us that he criticized Aristotle's *Topics* on minor grounds, saying that one topic (3.3 118b10–19) was too obvious, and rearranging Aristotle's order to make it easier to memorise. It is true that Alexander reports various more substantial points on which Theophrastus departed from Aristotle, but in his *Topics* at least he almost certainly accepted and built on Aristotle's foundations. We know nothing of Strato's *Introduction to* Topics (DL 5.59), but Van Ophuisen (163 and n. 248) notes that Alexander records an attempt by him to add a new topic to Aristotle's list, but in a work called *On the Prior and Posterior* (Fr. 30 W).

The greater part of this section consists of quotations given by Alexander, who probably did know Theophrastus' actual work. Later writers probably did not. On the *Topics* Theophrastus seems to have written only two books (**1** 159), though **1** 72—*the Reduction of Topics*—and **1** 269—*Preliminaries to the Topics*—may be relevant. See on **68** 18a and 19. Theophrastus did however write a number of other works on topics and on rhetoric which might have contained such material, but it is lost. We do however have names, mainly from DL, of several. As well as the *Topics* in two books (**68** 17)—rather a small compass to contain all that we might deduce from Averroes that it contained, there is a *Preliminaries to the Topics* in one book (**68** 19), an *On the art of Rhetoric* in one book, an *On kinds of Rhetorical arts* said to be in 61 books, where both the title and the number of books may be corrupt, and several which seem to be on individual aspects of rhetoric.

For a study of the Arabic account of topics see Ahmed Hasnawi, 'Topics and Analysis: the Arabic tradition' in R.W. Sharples (ed.) *Whose Aristotle? Whose Aristotelianism?* (Keeling series in Ancient philosophy) Aldershot: Ashgate 2001, 28–62.

32 Van Ophuijsen 1994. Slomkowski.

AB Alexander gives in two places almost identical accounts of Theophrastus' definition of a *topos.* We have no similar definition from Aristotle. We may conclude that these are Theophrastus' own words. We need not assume that they come from a *Commentary on Aristotle's* Topics, from which one might expect Alexander to have quoted more if it had existed. Other information is equally sparse, though the extent of Averroes' contribution is difficult to assess. Nor need we assume that the definition of a topic in **122AB** is unaristotelian. As Graeser 105–6 says, it is introduced by Alexander as if it were consistent with Aristotle.

On the other hand Eleonore Stump (Stump (1978) 159–78, 208–11) thinks that Theophrastus made some advances on Aristotle: she thinks that for Aristotle a topic was more a strategy than a principle, but Theophrastus introduced the term *parangelma*, for which we have used "precept" to distinguish the strategy from the principles under it. See now also De Rijk (2002) 628–36 for the meaning of *archai* in the *Topics*.

122A Alexander of Aphrodisias, *On Aristotle's* Topics 1, Introduction (*CAG* vol. 2.2 p. 5.21–7 Wallies) trans. *AlexTop* (VanOp) 7.

Slomkowsi 61–7.

In lines 6–7 we should note that Slomkowski (64 n. 96) records that in the Corrigenda et Agenda to his text of this passage Wallies prefers the plural προτάσεων ἐνδόξων to the singular we have printed. In line 7 for "probable" read "generally accepted" or "approved".

The quotation giving Theophrastus' definition of *topos* is almost identical with that given in **122B,** except for the addition of some comments, "turning our attention to it" in line 2, and "for either it includes the common and universal, which govern syllogisms, or at any rate from them such things can be demonstrated and grasped" (lines 3–5). The first of these seems unhelpful and is unlikely to be by Theophrastus, and the second is unclear and seems out of place. Both could be Alexander's insertions,[221] or could also be by a scholiast. Against the latter suggestion is the fact that the words are in the Suda: in its favour is the fact that the additions are obscure in sense and uncertain grammatically. It seems to me probable that the same source covers lines 6–7 as well, the first part explaining "limited in compass" and the *second* "unlimited with respect to individual cases". **122B** has a much more helpful comment.

For the difficulties in understanding lines 4–6 which we have noted in our footnote to the translation see now Slomkowski 67.

Averroes, *Middle Commentary on Aristotle's* Topics 2, prooemium (p. 525.5–7 Jéhamy)
Averroes quotes Theophrastus' definition of a topic before going on to Themistius' view.

---

[221] As is suggested by Stump 1978, 208.

122B Alexander of Aphrodisias, *On Aristotle's* Topics 1, Introduction (*CAG* vol. 2.2 p. 126.11–21 Wallies)

Slomkowski 61–7.

See Ebbesen vol. 1 110–12.

The quotations attributed to Theophrastus are identical, except that in **122A**, as we have seen, there is a kind of parenthesis the origin of which is unclear. It is not clear where the reference to Theophrastus ends in **122B**, and I give the following paragraphs as possibly by him (126.21–127.21)

"For if the enquiry were whether good is beneficial, starting from the topic before us we will take as a premise relevant to the problem before us 'If evil is harmful good is beneficial', (i.e. the one just mentioned). For from the topic before us this premise will derive both its existence and its credibility. And if we were asked whether the colour white is separative of sight, again we will start from the topic before us, and take the premise suitable and relevant to this problem: "If the colour black is compressive of sight, it is also the case that the colour white is separative of sight'.[222] But if, again, about pleasure, it were asked whether it is good, from the topic before us the premise will be taken which runs: 'If pain is evil, pleasure is good.' For all these, and similar ones, are included potentially, and without being defined in the topic before us.

"Again one topic about two: 'If what seems to belong more to something does not belong, neither would what seems less to belong belong to it.' This statement too is limited as regards the universal (for it shows that it is talking about what belongs more or less, just as the one before showed that it was about contraries) but it is unlimited with respect to individual cases; for it is not shown in it what belongs more to what. We can take 'more and less' with regard to colour and taste and goodness. It is indeed possible for those starting from the topic before us to be supplied with suitable premises about every problem in which 'more and less' can be used[223] For if it were asked, as it might be, whether wealth is a good or not, someone wishing to show that it is not a good, by

[222] Compare Theophrastus *On sensation* 86 and Simplicius *On Aristotle's* Categories 390.2–5.

[223] Albert the Great (our **315**.38–9) in a digression giving the views of Themistius and Theophrastus, says that they used the topic of (more and) less. For a discussion of other evidence see my commentary on Theophrastus' Psychology 172–3.

starting from the topic mentioned before, will take a premise relevant to the problem before him, by which he will produce a syllogism to show that it is not a good. For if health, which is more of a good thing than wealth, is not a good, neither would wealth be a good. Taking that premise, if a premise were added saying that health is not a good because, one might say, it turns out badly for some, and nothing good is the cause of evil, it would have been shown, in accordance with the topic before us, that wealth is not a good."

Cicero, *Topics* 7–8

Cicero attributes the invention of the *topos* to Aristotle and says that it is the seat (*sedes*) of an argument. For the relationship between Cicero and Theophrastus see Reinhardt (2003) 194–5.

Anonymous Segueriani *Art of rhetoric* (*RhGr.* Vol. 1 p. 382 Spengel-Hammer)
A *topos*, as Alexander says, is the starting-point of an epicheireme or a proof.

Aelius Theon, *Preliminary Exercises* 7 (*RhGr* vol. 2 p. 106 Spengel)
Theon's account of the *topos* is rather more elaborate than that of Alexander. For his methods see Reinhardt (2003) 195 n. 15.

Quintilian, Oratorical Education 5.10.20–2

In Fortenbaugh (2005) 177 there is the text of a papyrus from Newcastle, Hancock Mus. AREGYPT 522, mentioned as unpublished in our original collection of texts at **671**, but now published by Peter Parsons in *American Studies in Papyrology* 42 (2001) 153–65, which implies that Theophrastus wrote of 'common' and 'specific' topics. The context is rhetorical, but the distinction could apply also to dialectic.

123 Alexander of Aphrodisias, *On Aristotle's* Topics 2.2 109a34 (*CAG* vol. 2 p. 135.2–18 Wallies)

Van Ophuijsen 1994 138, 159–61. Stump 1978, 209–11. Slomkowski 62–3, 150–8.

Theophrastus seems to have introduced a new notion into topical theory, that of the *parangelma* or precept, which is the source of the topic as the topic is the source of the *epicheireme*.[224] Alexander uses *parangelma*

[224] See Van Ophuisen 151–3 esp. nn. 181–2. Slomkowski helpfully translates *parangelma* by "investigation-instruction".

at AlexTop 1 97.2 as a description of what Aristotle is doing at 106a1ff. giving advice on how to set about a problem. Van Ophuisen discusses the relations between *parangelma*, topic, and *epicheireme*. *Sustoicha*, meaning "coordinate items" here, has several senses in Aristotle and Theophrastus, and it is not entirely clear what is involved here. Alexander uses the term frequently at 104.1–9, and Van Ophuijsen in AlexTop (VanOp) translates it "members of the same set".

See now Fortenbaugh (2005) 66–9 for a discussion of rhetorical precepts, We can see here the overlap between rhetoric and dialectic.

4A Alexander of Aphrodisias, *On Aristotle's* Topics 1.6 102b27 (*CAG* vol. 2.2 p. 55.24–7 Wallies) trans. AlexTop (VanOp) 59

Van Ophuijsen 1994, 161–2, 167–8. Slomkowski 79–94.

At 1.5 101b37 Aristotle distinguishes four things,—definition (*horos*), property (*idion*), genus, and accident. For him there is a necessary connection between the essence of a thing and its definition, and again between its essence and its properties. Further, the genus is necessarily related to the species. Theophrastus seems to have tried to reorganise this, but how he did it is not clear. In this passage he appears to have put property and genus under definition, leaving accident separate, which seems a sensible thing to do, since accidents are outside definitions. But see **124B**.

This marks quite a departure from Aristotle, recognising only two divisions of subject—matter, but Van Ophuisen, 162 n. 239, points out that Aristotle himself (*Topics* 1.6 102b26–38) refers to this possibility and fears it would be less clear. Van Ophuisen (n. 246) also notes a similarity between Theophrastus' proposal and Eudemus' classification of questions (F. 25 W.)

Slomkowski has a careful study of *sumbebêkos* in Aristotle.

4B Proclus, *On Plato's* Parmenides 1, Introduction (col. 635.2–12 Cousin)

This differs from **124A** in that Theophrastus is said to have linked property and accident together. Property does occupy an intermediate position in Aristotle's scheme, being in some ways an aspect of things outside the definition, and so like an accident, but, because it is firmly linked to its owner, being also like a definition. Theophrastus can hardly have held both views at the same time, that accident and property are separate (**124A**) and that they go together, as here. Alexander is a reliable witness with a profound knowledge of Peripatetic logic, whereas Proclus had different interests and says very little about Theophrastus. On the other hand Proclus' account is detailed and plausible, and these views might have been held at different times. .

We may refer here to Boethius' *On Division* PL 892A[225] where he says that the later school of Peripatetics separated division *per se* from division *per accidens*, but the older ones (could) use an accident as a genus and accidents for species and differentiae. The older Peripatetics may include Theophrastus, but Magee 166–9 indicates the difficulties involved in interpreting this passage and is inclined to place even the earlier Peripatetics later than Theophrastus. .

125 Alexander of Aphrodisias *On Aristotle's* Topics 1.6 102a1 (*CAG* vol. 2.2 p. 45.10–13 Wallies) trans. AlexTop (VanOp) 48.

Before this Alexander has given an extended account of Aristotle's position in 102a1–17, and added that he also discusses the matter in the *PostA* and in the *Metaphysics*. He then says that Aristotle goes on to discuss the property, but inserts the present remark about Theophrastus, adding "and let this be treated as additional information from an outside source (ἐξώθεν προσιστορήσθω)" (Van Ophuijsen's translation).

But it is puzzling: the implication of the *gar* (for) in line 2 is that this was all that Alexander had to go on, and that he merely deduced from this one sentence that Theophrastus classed problems about sameness and the differentia with those about genus. But see on **123**, which suggests a more systematic approach, and on **127B**, where it seems likely that he had direct knowledge of some books of Theophrastus' *Topics*.

126 Averroes, *Middle Commentary on Aristotle's* Topics 3.3 118b10–19 Jéhamy

Gutas (1999) 129–130

Averroes' source here is Themistius's paraphrase of the *Prior Analytics*. (DG)

---

225 Now available in Anicii Manlii Severini Boethii *De divisione liber* Critical Edition, Translation, Prolegomena, & Commentary by John Magee Leiden: Brill 1998.

This looks like what we find at *Topics* 3.5 119a20–6, a discussion of addition and subtraction. The text to which it is attached, 118b10–19, seems less appropriate, but is fairly relevant. (PMH)

7A Simplicius, On Aristotle's Categories 11 14a19 (CAG vol. 8 p. 415.15–19 Kalbfleisch) trans. in Simplicius, *On Aristotle's* Categories 9–15 trans. Richard Gaskin London: Duckworth 2000 (*ACOA*) 60.

For Simplicius see on **112C**.

It is striking that Simplicius uses Theophrastus' *Topics* here rather than anything from his (putative) *Categories.* If that even existed, it was not used by Simplicius. As far as the *Topics* is concerned, even that may have been known only through Iamblichus.

We have printed the minimum necessary, but there is more in its context. Simplicius has begun (**414.27**?) with the views of Nicostratus (probably a Platonist of the second century AD) and of "the students of Porphyry", followed by "others" who give a peculiar and forced solution to the problem of explaining this passage of Aristotle. He then goes back in time to Theophrastus, criticizing him also. The passage continues:

On this Iamblichus says, "If there are ten genera of things that are, and all contraries are under these, either all will be ranged under one genus, or different under different, or they will be contrary in many ways, [226] each being placed under several genera.[227] Or else he (Theophrastus),[228] does not mean without qualification that contraries too are not in one genus, (and is) not referring to the categories as genera, but (he means) that there are no other genera of them which are contraries; for these are the principles of contraries, and under them the rest are ranged, since he himself too knew that excess and deficiency are ranged under one genus, quantity, from the point of view of being under one category, but as being contraries and the principles of contraries, he does not want them to be ranged under other contraries. Or else the primary contraries are not properly genera, and this is (the meaning of) "these

[226] Gaskin introduces "things said' before "in many ways', but that seems unnecessary.

[227] Gaskin closes the quotation from Iamblichus here, assigning what follows to Simplicius.

[228] When he says that contraries too are not in one genus.

too are not in one genus." Certainly the examples which they[229] use are such as to have in them prior and posterior."

The passage is brought in by Simplicius in connection with Aristotle's remark in *Categories* 14a19 that all opposites must be in the same genus, or in opposite genera, or themselves be genera. Aristotle gives examples, and concludes with—"good and bad are not in genera but are themselves genera of some things." Simplicius thinks Theophrastus' passage in his *Topics* has some connexion with this. It does not relate particularly to Aristotle's other passage in his own *Topics* (4.3 123b1–37) where there is no mention of "principles". Further, it may well have been taken over by Simplicius from Iamblichus' quoting of it. Iamblichus also made an attempt at explaining it.

Why does Iamblichus bring in the categories? It is not called for by Aristotle's passage, nor again by that of Theophrastus. Although Aristotle's passage is from the work now called the *Categories*, it is from that part known as the *Postpredicaments*. In both cases the examples given are irrelevant to categories. In lines 8–14 Simplicius seems to be rejecting Iamblichus' suggestion, and merely repeating what Theophrastus says. At 14–16 he tries again, but I find the text obscure.

The problem with *tauta* ("these") (line 3 of Greek) referring to the *archai* ("principles") is that this same sentence is taken up twice in what follows. In line 10 Simplicius actually substitutes *ta enantia* ("the opposites") for *tauta*, and in line 15 he suggests it means that *ta prota enantia* ("the primary opposites") are not in one genus. It looks as if Simplicius himself, or possibly Iamblichus, is puzzled about *tauta*. So in this passage we have various attempts at explaining it: a) (Iamblichus) trying to bring in categories; b) identifying *tauta* with *enantia* c) suggesting that the first contraries are involved. All this implies that Theophrastus' words were obscure.

I am grateful to Bob Sharples for a detailed consideration of this passage.

127B Alexander of Aphrodisias, *On Aristotle's* Metaphysics 5(D).10 1018a25 (*CAG* vol. 1 p. 381.11–15 Hayduck) trans. AlexMet (Dooley) 56.

Alexander is commenting on Aristotle's remarks about contraries. Here Theophrastus is mentioned together with Aristotle in the *Categories* and

[229] Presumably Aristotle and Theophrastus.

the *Topics*, but we can take it from **127A** that Theophtastus did mention motion and rest as well as good and bad. It is only Alexander who cites book 1 of Theophrastus' *Topics*. Elsewhere (**131**) he refers to Book 2, and it seems likely that he had direct access to it.

Aristotle, *Topics* 4.3 123b9.

Aristotle says that good is itself a genus.

-30 are all from Galen and connected with matters of genus and species. He mostly uses his knowledge of technical logic to attack other practitioners of whom he disapproves

28 Galen, *On the Method of Treatment* 1.3 (vol. 10 p. 26.11–16 Kühn)

Galen has previously referred to Plato's *Philebus, Sophist*, and *Politicus*, and Aristotle's *On the Parts of Animals.* All these are concerned with genera and differentiae. This passage tells us nothing about Theophrastus' methods, but they must have included what we can get from **124–9**.

Galen goes on to attack Thessalus for his way of classifying diseases.

9A Galen, *Against Lycus* 4.2 (*CMG* vol. 5.10.3 pp. 14.14–16 Wenkebach)

In line 1 we have put "also" instead of "at least" in the second impression.

Galen is attacking Lycus, a fellow physician and commentator on Hippocrates. The view supported by Theophrastus and the rest is that knowledge involves the ability to classify and differentiate. Galen goes on to taunt Lycus with his ignorance of the *Philebus* and then criticizes his treatment of liquids, and refers to Theophrastus' *On Flavour* or *On Juice* for which see on **384**.3 and **418** in Sharples' commentary on Biology.

9B Galen, *On the Method of Treatment* 2.7 (vol. 10 pp. 137.9–138.2 Kühn)

It has been suggested that *ouch' hena* would be better than *oudena* in line 3.

In his usual way Galen is attacking fellow doctors for their incompetence in logic. Aristotle's account is in *Metaphysics* 5.6 1016b31–5, where he distinguishes between being one in number and being one in species and so on, and Plato's at *Philebus* 14C, which is part of a long and tortuous discussion of the one and many problem.

Galen, *On the Method of Treatment* 2.7 p. 142.2–4.

Galen refers to Aristotle, Theophrastus and Plato about the difference between being one in species and being one in number.

130 Galen, *On the Method of Treatment* 1.3 (vol. 10 pp. 19.13–17 and 22.5–16 Kühn)

Galen is attacking Thessalus in extravagant terms. For us the only important point is that he ranks Theophrastus with Plato, Aristotle and Chrysippus as worthy to be judges of Thessalus" arguments.

Galen, *On the Method of Treatment* 1.3 p. 28.2–4.

Galen criticizes Thessalus for, among other things, attacking Theophrastus and the Stoics on dialectic.

131 Alexander of Aphrodisias, *On Aristotle's* Metaphysics 5(D).15 1021a31 (*CAG* vol. 1 p. 408.18–22 Hayduck) trans. AlexMet (Dooley) 90.

This is presented as an actual quotation from Theophrastus, drawn from his *Topics* Book 2. Repici (fr. 61) has interpreted the first sentence slightly differently, supplying τὰ πρός τι ("relatives") where we have "harmony". "Harmony" or "an orderly arrangement of parts" is mentioned several times by Alexander in the preceding lines and either interpretation is possible. Dooley has: "the fact that [some relatives] are referred to the things in which they are [is confirmed] by what Theophrastus says...". Alexander is commenting on Aristotle's chapter on relatives. He follows our passage with a quotation from Aristotle's own *Topics.* The main point of Theophrastus' remarks appears to be that there are many kinds of relatives, and every case must be looked at separately. His examples symmetry, disposition and synthesis, repeat two of the three given by Aristotle, symmetry and disposition (*hexis*), but where Aristotle has "state" (*diathesis*) Theophrastus has "synthesis". But his other examples, memory and sleep and thought (*hupolêpsis*), are not to be found in Aristotle in a similar context.

Aristotle, *Topics* 4.4 125a33–5 (This is the correct reference).

Aristotle's words clearly lie behind the first part of those of Theophrastus. He uses two of the three examples, but has *diathesis* where Theophrastus gives *sunthesis*. And Theophrastus' remarks about memory and the rest are not in Aristotle.

Appendix 2 *Florentine fragment on Topics*

Philippson, R. 'Il frammento logica Fiorentina' *Rivista di Filologia e d'Istruzione Classica* VII (1929) 495–506. Solmsen, F. Il frammento logica Fiorentina' ibid. 507–10. Van Ophuisen, *Peripatetic Rhetoric after Aristotle* (RUSCH 6) New Brunswick Transaction Publishers 1994 158. See also 163.

This is clearly a part of a long work, as it refers back to the thirtieth topic. Further, it is closely related to Aristotle's *Topics* 5.8 138a30–b26. It is shorter and omits the examples that Aristotle gives, but instead makes use of letters/numerals. It also differs in the order of the topics it covers, in a way similar to the differences between Theophrastus and Alexander over the order of totally hypothetical syllogisms. The final lines refer back to the thirtieth topic, concerning things which were related similarly by analogy or equality. This may refer to something similar to Aristotle's *Topics* 5.7 136b33 which also uses the pression *homoiôs echontôn* (found in our lines 7 and 10); further, at 138b24 there is the expression "by analogy" found in our line 10.

There is no positive reason to assign this to Theophrastus, but Barnes (1985) 134–5 suggests that it might be part of *The Reduction of Topics* (**68** no. 18a).

It would be risky to refer to the possibility of Theophrastus having been like the Stoics in using numerals. This usage is different from that of the Stoics, who used numerals only for propositional variables, and Aristotle himself (*PA* 1.15 34a5–34 2.2 53b12–15) used *to A* etc. as propositional variables.[230] Walter Cockle says that it is unusual for letters to be used in this way in pagan literary texts. It is common in documentary papyri, and also in unsophisticated Christian texts.

32 Averroes, *Middle Commentary on Aristotle's* Topics 6.5–14 (pp. 602.14–603.6, 603.23–604.2, 620.7–8 Jéhamy)

Gutas (1999) 129–130

Averroes' source here is Themistius's paraphrase of the *Prior Analytics*. (DG)

[230] Sharples raises *PA* 2.2 53b16–17 as a problem. In its context, this one statement is odd.

This is part of a much longer section of Averroes' Commentary, in which Theophrastus' name occurs three times, and which arguably may all be based on him. Averroes says that in the long list of topics which he proceeds to give in connection with *Topics* 6 he is following the order of Themistius and before him Theophrastus, because this order is thought to be technically superior and easier to remember—suggesting that the list of topics were learned by heart.

In fact the list follows Aristotle quite closely both in order and in content, but there are differences, suggesting that Theophrastus was trying to improve on his master. The list omits topic 35 and puts 40 between 43 and 44. (PMH)

133 Alexander of Aphrodisias *On Aristotle's* Topics 2.3 110b16 (CAG vol. 2.2 p. 154.16–23 Wallies)

Hintikka 1973 12 says that in all Aristotle there is no trace of a difference between *pollachôs legetai, pleonachôs legetai* and *posachôs legetai* ("it is said in many ways, it is said in several ways" and "it is said in how many ways"). Often they are used interchangeably. Galen, *De Captionibus* ("*On Sophisms based on Language*"), does not mention Theophrastus, but treats Aristotle as the only earlier commentator on ambiguity worth taking seriously. Galen does refer to Eudemus, on whose *Peri Lexeos* ("*On Expression*") he wrote 3 books, and Edlow (1977 54–5) suggests that Eudemus wrote about apparent ambiguity. Graeser 56 (his XV) suggests the work might be like Aristotle's *Metaphysics* Delta.

At Barnes' suggestion we have added *cf. Aristotelis Anal. Post.* 1.1 71a25–b8 to the apparatus for 1–9.

This passage in the first chapter of the *PostA* discusses a problem about knowledge, using triangle as an example. It shows that there is more than one way of having knowledge.

134 Alexander of Aphrodisias, *On Aristotle's* Topics 5.2 130a8 (CAG vol. 2.2 pp. 378.24–379.3 Wallies)

Theophrastus gives examples of ambiguity, both with single words and with groups of words. First are ambiguous groups, of which he gives a) ἐπίστασθαι γράμματα. This can be rendered as "to know" and "letters". The Greek words are said to have a single meaning each, but to produce ambiguity when put together. This example is taken from Aristotle's

Sophistic refutations 4 166a18–21, where it is said that it can mean either that letters themselves have knowledge, though that is not very plausible in this case, or that another knows letters. b) In “Dion Plato beats” the order in English is unusual and one might say poetical, and the Greek depends on the words being in reported speech so that either name can be the subject, with the other being the object of the verb. See also on **88**.

In line 8 in the second impression we have written “generally accepted” for “probable”. We should have done the same in line 6.

135 Alexander of Aphrodisias, *On Aristotle’s* Topics 1.2 101a26 (*CAG* vol. 2.2 p. 27.11–18 Wallies) trans. AlexTop 30–1.

Moraux (1968) 301. Van Ophuijsen (1994) 150–3.

. . . This passage tells us more about Alexander’s methods of arguing than it does about those of the early Peripatetics, but it does suggest that the books of theses by Aristotle and Eudemus were still in existence in Alexander’s time. See also on **74** and **89**.

136 Anonymous, *On Aristotle’s* Posterior Analytics 2.15 98a24 (*CAG* vol. 13.3 p. 590.4–10)

Wallies prints this after works by Philoponus. It is found in five MSS, but the text is in a bad state.

Prantl 350 asks whether this refers to a separate work or is just on *PostA* 2.14–18. Moraux, 1979 on 2.131–46 thinks it may be from Alexander.

After this passage there is an account of what Aristotle says.

The main point is to distinguish between Aristotle’s method in the *PostA* and that of Theophrastus in an unnamed work. There is nothing in our list of titles that stands out as an obvious candidate. We may compare **97**, where two different types of work by Theophrastus on the same subject-matter are mentioned.

A further trace of Theophrastus is perhaps to be found in Cicero’s *Topics* 83 which refers to *descriptio* as a separate way of explaining something. Not only does he equate it with the Greek word *charakter*, but his examples are: what the miser and the flatterer are like, and he adds: and other things of that kind in which a man’s nature and life is described. This is given as a fifth way of finding out what something is;

the first four are familiar logical concepts, notion, property, division and partition, but *descriptio* both seems different and is introduced as such by Cicero. Since both the language and the examples recall Theophrastus, it is reasonable to think of him here, though it would be going too far to see his actual *Characters* as written primarily for the purposes to which Cicero refers.

The notion that topics are useful for finding out arguments is to be found in Cicero in various colourful forms: at *Topics* 7 it is that of hunting out hidden things when their hiding-place is known. Not merely the words, but Cicero's whole method, involves treating topics in this way. Alexander does not say that Theophrastus made this point in his *Topics*, which, as we have seen, was closely connected with Aristotle's *Topics*, but other evidence suggests that he did It is worth noting, however, that at 55 Cicero uses the baffling word *enthumeme*, which does not appear to be a Stoic term. Since however he says that he is giving a special sense used by rhetors we need not bring in the Peripatetics here. Cicero's *Partitiones oratoriae* provides further material: at 41 *descriptiones* are frequently used for Aristotle's purposes of definition, and at 65 a *descriptio* is like a picture to be portrayed, where the Latin *exprimenda* seems to involve the equivalent of *charakter*, and where Aristotle's examples are the miser (again) and the proud. See Moraux (1968) 301–4 for Cicero on Peripatetic dialectic.

BIBLIOGRAPHY

See also the Abbreviations on p. xi

Ackrill, J. Aristotle's *Categories* and *de Interpretatione*, Oxford: Clarendon Press 1963.

Alexander of Aphrodisias, *On Aristotle's* Prior Analytics 1–7 trans. J. Barnes, S. Bobzien, K. Flannery and K. Ierodiakonou, Ithaca N.Y.: Cornell University Press, 1991 *ACOA*.

——. *On Aristotle's* Prior Analytics 8–13 etc. trans. Ian Mueller with Josiah Gould, London: Duckworth, 1999a (*ACOA*).

——. *On Aristotle's* Prior Analytics 1.14–22 trans. Ian Mueller with Josiah Gould, London: Duckworth, 1999b (*ACOA*).

——. *On Aristotle's* Prior Analytics 1.23–31 trans. Ian Mueller, London: Duckworth 2006a (*ACOA*).

——. *On Aristotle's* Prior Analytics 1.32–46 trans. Ian Mueller, London: Duckworth 2006b (*ACOA*).

——. *On Aristotle's Topics* 1 trans. J.M. Van Ophuijsen, London: Duckworth 2001 (*ACOA*).

——. *On Aristotle's Metaphysics 4* trans. Arthur Madigan, London: Duckworth, 1993 (*ACOA*).

——. *On Aristotle's Metaphysics 5* trans. W.E. Dooley, London: Duckworth, 1993 (*ACOA*).

Ammonius, *On Aristotle's* On Interpretation 1–8 trans. David Blank, London: Duckworth, 1996 (*ACOA*).

——. *On Aristotle's* De interpretatione (cod. Paris. Gr. 2064 Tarán), Meisenheim am Glan: Verlag Anton Hain, 1978 (*Beiträge zur klassischen Philologie* 95).

Aristotle, De interpretatione, C.W.A. Whitaker Oxford: Clarendon Press, 1996.

Averroes, *Middle Commentary on Aristotle's* Prior Analytics, ed. G. Jéhamy, *Kitāb al-Qiyās*, in Ibn Rušd Talhīs Mantiq Aristū, vol. 1, Beirut 1982.

—— ed. M.M. Kassem, C.E. Butterworth, and A.A. Haridi, Cairo 1983.

——. *Middle Commentary on Aristotle's* Topics, ed. G. Jéhamy, *Kitāb al-Gadal*, in Ibn Rušd. *Talhīs Mantiq Aristū*, vol. 2, Beirut 1982, 497–666

——. *Questions*, ed. Gamāladdīn al-ʿAlawī, Maqālāt fī l-mantiq wa-l-ʿilm at-tabīʿī li-Abī l-Walīd Ibn Rušd, Casablanca 1983.

Badawi, Abdurrahmān *Aristū ʿinda l-ʿArab*, Cairo 1947.

—— *La transmission de la philosophie grecque au monde arabe*, Paris 1968

—— *Šurūh ʿalā Aristū mafqūda fī l-yūnāniyya wa-rasāʾil uhrā*, Beirut 1971.

Baltussen, H. *Theophrastus against the Presocratics and Plato*, Leiden: Brill, 2000.

Barnes, J. 'Aristotle's Theory of Demonstration' *Phronesis* 14 (1969) 123–52, revised in *Articles on Aristotle* (ed. Barnes, Schofield and Sorabji) London: Duckworth, 1975a 65–67.

——. Aristotle, *Posterior Analytics*, Oxford: Clarendon Press 1975b, 2nd ed. 1993

——. "Terms and Sentences: Theophrastus on Hypothetical Syllogisms" *Proc. Of the Britisn Academy* 69 (1983) 279–362.

——. 'Theophrastus and Hypothetical Syllogistic' in Wiesner, J. (ed.) 1987 557–76 and in Fortenbaugh W.W. with Huby, P.M. and Long, A.A. 1985 (*RUSCH* 2).

——. 'Grammar on Aristotle's Terms' in Frede and Striker (1996) 175–202.

——. 'Aristotle and Stoic Logic' in K. Ierodiakonou (1999) 23–53.

——. 'Syllogistic in the Anon Heiberg' in K. Ierodiakonou (2002) 97–137.

Becker, A. *Die aristotelische Theorie der Möglichkeitsschlusse*, Berlin: Junker und Dünnhaupt 1933.

Blumenthal, H.J. *Aristotle and Neoplatonism in Late Antiquity*, London: Duckworth 1996.
Blumenthal, H.J. and Robinson, H. *Aristotle and the Later Tradition* Oxford 1991 (*OSAP* suppl. vol.)
Bobzien, Susanne 'Why the Order of the Figures of the Hypothetical Syllogism was Changed' *CQ* ns 50 (2000).
—— 'The Development of *Modus Ponens* in Antiquity' *Phronesis* 47 (2002a).
—— 'Pre-Stoic Hypothetical Syllogistic in Galen's *Institutio Logistica*' in Nutton (2002b).
Bocheński, I.M. *La Logique de Théophraste*, Fribourg: Librairie de l'Université 1947 (*Collectanea Friburgensia*, N.S. Fasc. 32).
——. *Ancient Formal Logic*, Amsterdam: North Holland 1951.
Bodnár, I. and Fortenbaugh, W.W. *Eudemus of Rhodes*, New Brunswick (USA) and London (UK): Transaction 2002 (*RUSCH* 11).
Brunschwig, J. *Aristotle Topiques I Livres 1–4*. Paris: Budé 1967.
—— '"*Indéterminé*" et "indéfini" dans la logique de Théophraste' *Revue philosophique de la France et de l'Etranger* 172 (1982) 359–70.
Burnyeat, M. 'Enthymeme: Aristotle on the Logic of Persuasion' in D.J. Furley and Alexander Nehemas *Aristotle's Rhetoric: Philosophical Essays* Princeton University Press 1994 *(XIII Symposium aristotelicum).*
Courcelle, P. *Late Latin Writers and their Greek Sources* trans. H.E. Wedeck Cambridge, Mass.: Harvard University Press 1969.
Dürr, K., *The Propositiomal Logic of Boethius*, tr. N. Martin Amsterdam: North Holland 1951.
Ebbesen, Sten *Commentators and Commentaries on Aristotle's* Sophistici Elenchi: a study of post-Aristotelian ancient and medieval writings on fallacies Leiden: Brill 1981.
Edlow, R.B. *Galen on Language and Ambiguity*, Leiden: Brill 1977.
al-Fārābī, *On Aristotle's* De Interpretatione, ed. W. Kutsch and S. Marrow, *Šarh al-Fārābī li-Kitāb Aristūtālīs fī l-'Ibāra*, Beyrouth 1971
Flannery, K.L. *Ways into the Logic of Alexander of Aphrodisias*, Leiden: Brill 1995.
Fortenbaugh, W.W. 'Theophrastean titles and Book Numbers: Some Reflections on Titles Relating to Rhetoric and Poetics' in *Fragmentsammlungen Philosophischer Texte der Antike/Le raccolte dei frammenti di filosofi antichi = Aporemata* 3, ed. W. Burkert *et al.* Göttingen: Vandenhoeck & Ruprecht 1998 and in Fortenbaugh 2003.
——. 'Teofrasto di Ereso: Argomentazione retorica e sillogistica ipotetica', *Aevum* 74 (2000), and as 'Theophrastus of Eresus: Rhetorical Argument and Hypothetical syllogistic' in Fortenbaugh (2003).
——. 'Eudemus' Work *On Expression*' in Bodnár and Fortenbaugh (2002) 59–83.
——. *Theophrastean Studies*, Stuttgart: Franz Steiner Verlag 2003 (*Philosophie Der Antike* 17).
——. *Theophrastus of Eresus Sources for his Life, Writings, Thought & Influence* Commentary vol. 8 Sources on Rhetoric and Poetics Leiden: Brill 2005.
Fortenbaugh, W.W. with Huby, P.M. and Long, A.A. *Theophrastus of Eresus* On his Life and Work, New Brunswick: Transaction 1985 (*RUSCH* 2).
Fortenbaugh W.W. and Mirhady, D.C. *Peripatetic Rhetoric after Aristotle*, New Brunswick: Transaction 1994 (*RUSCH* 6).
Fortenbaugh, W.W. and Wöhrle, G., *On the Opuscula of Theophrastus* Stuttgart: Franz Steiner Verlag 2002 (*Philosophie der Antike* 14)
Frede, M. *Essays in Ancient Philosophy*, Minneapolis: University of Minnesota Press 1987.
Frede, M. and Striker, G. *Rationality in Greek Thought*, Oxford: Clarendon Press 1996.
Gottschalk H.B. 'Aristotelian philosophy in the Roman world from the time of Cicero to the end of the second century AD' *ANRW* 2 36.2 (1987).
Graeser, A. *Die logischen Fragmente des Theophraste*, Berlin/New York: De Gruyter 1973.
Gutas, D. 'The Life, Works, and Sayings of Theophrastus in the Arabic Tradition', in Fortenbaugh with Huby and Long 1985, 63–96; repr. in Gutas (2000), no. VII.
——. *Avicenna and the Aristotelian Tradition*, Leiden: Brill, 1988.

——. 'Averroes on Theophrastus, through Themistius', in Gerhard Endress and Jan A. Aertsen (eds.), *Averroes and the Aristotelian Tradition*, Leiden: Brill, 1999, 125–144.

——. *Greek Philosophers in the Arabic Tradition, Aldershot*: Variorum 2000.

Hadot, I. *Simplicius. sur les catégories* vol. 1. Leiden: Brill 1990.

Hein, Christel *Definition und Einteilung def Philosophie von spaetentiken Einleitungsliteratur zur arabischen Enzyklopaedie* Frankfurt/Bern/New York: Peter Lang 1985.

Hintikka, K.J.K., 'Aristotle's different Possibilities' in Moravscik J.E.M. *Aristotle*, London: Macmillan 1967/8 and in Hintikka (1973).

——. *Time and Necessity* Oxford: Clarendon Press 1973.

Huby, P.M., 'Did Aristotle Reply to Eudemus and Theophrastus on Some Logical Matters?' in Bodnár and Fortenbaugh (2002).

Ibn-Abī-Usaybi'a, *'Uyūn al-anbā' fī tabaqāt al-atibbā'*, ed. by August Müller, 2 vols, Cairo-Königsberg 1882–1884.

Ibn-an-Nadīm, Kitāb al-Fihrist, Gustav Flügel, Johannes Roediger, August Müller, eds. 2 volumes, Leipzig 1871–2.

Ierodiakonou, Katerina, *Topics in Stoic Philosophy* Oxford: Clarendon Press 1999.

——. *Byzantine Philosophy and its Ancient Sources* Oxford: Oxford University Press 2002.

Kieffer, J.S. Galen's *Institutio Logica*, English Translation, Introduction, and Commentary, Baltimore: Johns Hopkins 1964.

Kneale, W.C. and M. *The Development of Logic*, Oxford: Clarendon Press 1962.

Lameer, J. *Al-farabi and Aristotelian Syllogistic* Leiden: Brill 1994.

Lee, T.-S. *Die griechische Tradition der aristotelische Syllogistik in der Spätantike*, Göttingen: Vandenhoeck & Ruprecht 1984 (*Hypomnemata* 79).

Lejewski, C. 'On prosleptic syllogisms' *Notre Dame Journal of Formal Logic* II (1961) 158–76.

——. 'On prosleptic premisses' *Notre Dame Journal of Formal Logic* XVII (1976) 1–18.

Lukasiewicz, J. *Aristotle's Syllogistic* Oxford: Clarendon Press first ed. 1951, second ed. 1957.

Maier, H. *Die Syllogistik des Aristoteles*, Tübingen: H Laupp'schen Buchhandlung 1896–1900, repr. Leipzig: K.F. Koehler 1936.

McCall, S. *Aristotle's Modal Syllogisms* Amsterdam: North Holland Publishing Company 1963.

Mignucci, M. 'Per una nuova interpretazione della logica modale di Teofrasto', *Vichiana* 2 (1965) 3–53.

——. Aristotele, *Gli Analitici Primi* (Italian translation with introduction and Commentary), Naples: Luigi Loffredo 1969.

——. 'Ammonius on Future Contingent Propositions' in Frede and Striker (1996).

——. 'Theophrastus' Logic' in Van Ophuijsen and van Raalte (1998) 39–65.

——. 'La Critica di Teofrasto alla logica Aristotelica' in C. Natali and S. Maso *Antiaristotelismo*, Amsterdam: Hakkert 1999 21–39

Mohaghegh, M. and Izutsu, T. *Collected Texts and Papers on Logic and Language* Tehran 1974 (*Wisdom of Persia VIII*).

Moraux, P. 'La joute dialectique d'après l'huitième livre des *Topiques*' in Owen (ed.) 1968.

——. 'La critique d'authenticité chez les commentateurs grecs d'Aristote' in *Mélanges Mansel* Ankara: Türk Tarih Kurumu Basimevi 1974, 264–88.

——. *Le commentaire d'Alexandre d'Aphrodise aux 'Secondes analytiques' d'Aristote* (*Peripatoi* 13) Berlin: De Gruyter 1979

——. *Der Aristotelismus bei den Griechen von Andronikus bis Alexander von Aphrodisias* vol.II "Der Aristotelismus im I und II Jh. n. Chr." Berlin: De Gruyter 1984 (*Peripatoi* 6).

Müller, August *Die griechischen Philosophen in der arabischen Überlieferung* [Festschrift Bernhardy], Halle 1872.

Nutton, V. *The Unknown Galen BICS* Suppl. 77, London: Institute of Classical Studies University of London 2002. Johannes

Ibn-an-Nadīm, Kitāb al-Fihrist, Gustav Flügel, Johannes Roediger, August Müller, eds, 2 volumes, Leipzig 1871–2.

Owen, G.E.L. (ed.) *Aristotle on Dialectic: the Topics* Oxford: Clarendon Press 1968 (*Proceedings of the Third Symposium Aristotelicum*).

Patterson, R. *Aristotle's Modal Logic* Cambridge: Cambridge University Press 1995.

Patzig, G. *Aristotle's Theory of the Syllogism*, Dordrecht: D.Reidel 1963.

Peters, F.E. *Aristoteles Arabus*, Leiden: Brill 1968.

Prantl, C. *Geschichte der Logik im Abendlande.* Leipzig 1858 repr. 1927 Buchhandlung Gustav Fock.

Prior, A.N. *Formal Logic* second edition Oxford: Clarendon Press 1962.

Repici: La logica di Teofrasto Studio critico e raccolta dei frammenti e delle testimonianze Bologna: Il Mulino 1977.

Rescher, N. 'Averroes' Quaesitum on Assertoric (Absolute) Propositions', *Journal of the History of Philosophy* 1 (1963), repr. in his *Studies in the History of Arabic Logic*, 91–105.

——. *Studies in the History of Arabic Logic*, Pittsburgh 1963.

——. *Temporal Modalities in Arabic Logic* Foundations of Language Suppl. Series Dordrecht: D. Reidel 1967

——. *Studies in Modality, American Philosophical Quarterly* Monograph 8 Oxford: Basil Blackwell 1974.

——. 'Modal Syllogistic in Arabic Philosophy' in Rescher 1974 17–56

Rijk, L.M. de *Logica Modernorum* vol. I 1962 vol II 1967 Assen: Van Gorcum

——. 'On the Chronology of Boethius' Works on Logic' *Vivarium* 2 (1964) 1–49, 122–62

——. *Aristotle Semantics and Ontology* vol. I *General Introduction to. The Works on Logic* Leiden: Brill 2002

Rose, Lynn *Aristotle's Syllogistic*, Springfield, IL: Thomas 1968.

Rosenberg, S. and Manekin, C. 'Themistius on Modal Logic. Excerpts from a Commentary on the *Prior Analytics* attributed to Themistius', *Jerusalem Studies in Arabic and Islam* 11 (1988) 83–103

Ross, D. *Aristotle's Prior and Posterior Analytics* Oxford: Clarendon Press 1957.

Schenkeveld, Dirk 'Theophrastus' Rhetorical Works: One Rhetorical Fragment the Less, One Logical Fragment the More' in Van Ophuijsen and van Raalte (1998) 67–80.

David Sedley 'The Stoic-Platonist Debate on *kathêkonta*' in Ierodiakonou (1999).

Sharples, R.W. 'Temporally qualified necessity and impossibility' *LCM* 3 (1978) 89–91.

Shiel, J. 'Boethius' Commentaries on Aristotle' *Medieval and Renaissance Studies* 4 (1958) 217–44; repr. with Postscript in Fuhrmann, M. and Gruber, J. (eds.) *Boethius* Darmstadt (1984) 155–83 (*Wege der Forschung* 583) and (revised) in Sorabji (1990) 349–72.

Slomkowski, P. *Aristotle's* Topics, Leyden: Brill 1997.

Sollenberger, Michael 'Diogenes Laertius 5.36–57 The *Vita Theophrasti*' in Fortenbaugh with Huby and Long (eds.) (1985).

Solmsen, F. 'Dialectic without the Forms' in Owen (ed.) (1968).

Sorabji, R. *Aristotle Transformed*, London: Duckworth 1990

Speca, A. *Hypothetical Syllogisms and Stoic Logic*, Leiden: Brill 2001.

Stern, S.M., Hourani, A. and Brown, V. *Islamic Philosophy and the Classical Tradition*, London: Cassirer 1972.

Striker, G.'Perfection and Reduction in Aristotle's *Prior* Analytics' in Frede and Striker (1996) 203–19.

Stump, E. Boethius' *De topicis differentiis*, Ithaca and London: Cornell University Press 1978.

Sullivan, Mark W. *Apuleian Logic* Amsterdam: North Holland 1967.

Van Ophuijsen, J. 'Where Have the Topics Gone?' in Fortenbaugh and Mirhady (1994) 131–73.
Van Ophuijsen, J. and van Raalte, M. *Theophrastus Reappraising the Sources* Leiden: Brill 1998 (*RUSCH* 8).
Wartelle, A. *Inventaire des Manuscrits grecs d'Aristote et de ses Commentateurs*, Paris: Les Belles Lettres 1963.
Wehrli, F. *Die Schule des Aristoteles: Texte und Kommentar* 10 vols. + 2 suppls. 2nd ed. Basel: Schwabe 1967–78.
Westerink, L.G. 'Elias on the Prior Analytics' *Mnemosyne* 14 (1961).
Whitaker, C.W.A. Aristotle's *De interpretatione*, Oxford: Clarendon Press 1996.
White, S. 'Opuscula and Opera in Theophrastus' Works' in Fortenbaugh and Wöhrle (2002)
Wiesner, J. (ed.) *Aristoteles Werk und Wirkung*—Paul Moraux gewidmet, Berlin: De Gruyter 1987.
Wildberg, C. 'Three Neoplatonic Introductions' *Hermathena* 149 (1990) 33–51.
Wilson, N.G. *Scholars of Byzantium*, Baltimore: The Johns Hopkins University Press 1983.
Zeyl, D.J. *Encyclopedia of Classical Philosophy* Westport, Conn.: Greenwood Press 1997.
Zimmermann, F.W. *Al-Farabi's Commentary and Short Treatise on Aristotle's De Interpretatione*, Oxford: Oxford University Press, 1981.

INDICES TO THE TEXTS

Principal Terms: Greek

Principal Terms: Latin

possibilis possible **101.**8,11,14,15,16,31,33
possibilitas possibility **101**.4bis,20, 21,22,24,27bis,29bis,36,37
praedicare predicate **91D.**21,23,52, 53,55,62,63,64?,67,70,79, 80,83bis,84
principium starting point **91D.**13
proferre produce **118.**3,6,9
propositio proposition **82B.**4 **91D.**16,20,24,26 **92.**3 **95A.**9 **101.**19,22,23 **113D.**3,7,8
proprietas property **82B.**2,7

redarguere argue to the contrary **113D.**1
reflexim by conversion **95A**.2
refractio reflexion **91D.**15,34
respicere look back at **101**.21
rigor force **101.**9

sequi be consequence of **101.**5,22,24, 25,26,28,29,30,36,39 follow **101.**20
significare signify **118.**9
significatio meaning **101**.35
singularitas individuality **82B.**2
singulus individual **82B.**2
species aspect **101**.33
subiectivus subjective **95A.**10
subiectum subject **118**.6
substantia substance **118.**4,8
syllogismus syllogism **91D.**13,28 **91E.**1 **111A.**4,5 **113D.**2,7,8,9

terminus term **91D.**19bis,21bis,22bis, 29ter,38bis,39bis,40, 41bis,51bis,52bis, 53,54,55,62bis,63bis,64bis,66,67bis,68, 69bis,78ter,79,80bis,82,83bis,84
tollere?? destroy **101.**15

universalis universal **82B.**6 **91D**.16,20,28,36bis,37,40,49,61bis, 65,68,73,82 **95A.**2
universaliter as a universal **91D.**34

vis force **101**.7,18.31.35 strength **101.**23
vocabulum name **82B.**5

Principal Terms: Arabic

(The order of the entries follows the Arabic alphabet. When a fragment number is not followed by a line number then the word in question occurs repeatedly in that fragment.)

ʾBD	
abadī	eternal 100C.7
ʾLF	
allafa	to combine 94.23; 106I.6
taʾlīf	combination, composition 94.17–30; 111C.1
muʾallaf	combined 98C.9; 106H.28
BRHN	
burhān	demonstration 91C.1; 98B.20
BSṬ	
basīṭ	simple, absolute 87F.1
BLĠ	
balāġa	rhetoric 98B.20
ǦDL	
ǧadal	dialectic 98B.20
ǦZʾ	
ǧuzʾ	part 106H
ǧuzʾī	particular 105.37
ǦNS	
ǧins	genus 81B.4; 98B.10; 98E.1; 132.2–3,14–17
ǦWZ	
ǧawwaza	to allow 98A.6
ḤǦǦ	
ḥuǧǧa	argument 106I.6
iḥtaǧǧa	to argue 98B.17,39–40; 105.8,16; 106H.24,27
iḥtiǧāǧ	argumentation 98B.16,20
ḤDD	
ḥadd	term 98B.30; 106H. 14–16, 20 / definition 99.5; 132
taḥdīd	defining 121.1
ḤKM	
ḥukm	statement, declaration 98A.2,7
ḤML	
ḥamala	to predicate 94.7–8,32; 105.34; 106H
ḥaml	predication 94; 105.25,29, 34–36; 106H.6–8,22
maḥmūl	predicate 98A.16; 98B.13; 98C.7; 98D.2; 98G.7; 100C.6; 105.34
ḤWL	
muḥāl	absurd, inconsistent 99.4
ḪLF	
bi-ṭarīq al-ḫulf	reductio ad impossibile 105.7
DWM	
dāʾiman	always 98A.10,14
lā dāʾiman	not always 98A.4,14
dawām	duration 98A.17
al-lā-dawām	non-duration 98A.17
D̲W	
min d̲ātihi	of or by itself 94.29
RDD	
radd	reduction 94.30
RSL	
mursal	simple 98B.27
RWQ	
riwāqī	Stoic 111C.3
aṣḥāb ar-riwāq	Stoics 111C.2
SʾL	
masʾala	problem 94.13
SLB	
salb	negation 106H.25
sālib	negative 81B.8; 87F.1; 94.20; 98A.6, 8; 106H.26
SLM	
sallama	to admit, to grant 94.26–27; 105.36, 38
taslīm	admission 94.24, 28–29

qaṣd awwal	primary intention 81B.6
maqṣūd	what is intended 98F.3
QḌY	
qaḍiyya	proposition 98E.2
QWL	
maqūla	category 71D; 132.6,9
QYS	
miqyās	syllogism 98B.28; 98C.9; 105; 111C.5–6
qiyās	syllogism 81B2; 91C.1; 94.28,33; 98B.29; 105.9,19; 106H.9; 106I.6,9; 111C.4
muqāyasa	comparison 126.4
KBR	
al-kubrā	major (term) 98B.22; 105; 106H; 106I.2,9
KḎB	
kāḏib	false 98B.31
kaḏib	untruth, falsehood 98B.24
KLL	
kull	whole 106H
fī l-kull	universally 98B.26
kullī	universal 98A; 98B.18,23; 105.36
kulliyya	totality 105.28
KM	
kammiyya	quantity 105.28
KYF	
kayfiyya	quality 105.29
LZM	
ilzām	sequence 105.36
MDD	
mādda	matter (of syllogism) 98B.8,11; 105.17–18; 106H.27; 106I.6
MŠY	
maššā'	Peripatetic 94.2; 98B.11; 105.4,10,22; 106H.1
MKN	
imkān	possibility 98A.2; 98B.8;99
bi-imkān	possibly 98C.7; 98G.8
mumkin	possible 98B.9–10, 13, 21; 98C.8; 98D.3, 5; 98F; 98G.2; 99.3,6; 105
NTǦ	
antaǧa	to yield a conclusion 94.27; 98B.24,30–31,35; 106I.10
intāǧ	derivation of conclusion 105.17–18
natīǧa	conclusion 91C.2; 98C.10; 105; 106H; 106I
NQḌ	
nāqaḍa	to dispute 94.27
NQL	
naqala	to transform 94.15; 98A.7 / to transfer 100C.6
WǦB	
īǧāb	affirmation 106H.25
mūǧib	affirmative 81B.8; 87F.1; 98A.6–7; 105.5; 106H.26
WǦD	
wuǧida	to exist 98B.13.18; 98C.3; 106H.27; 106I.7; 111C.6
wuǧūd	existence 98A.12; 98B.18,45; 98F.2–3; 98G; 99.1; 106H.3
wuǧūdī	existential 98B; 98C.6; 98D; 98F.1; 98G.1; 105
mawǧūd	existent 98A.3; 98B.26,28,43; 98C; 98D.3, 5; 98F.4; 106H.13 / belong to 98D.2; 98G.8; 99.4,6; 100C.4
WǦH	
ǧiha	mode, modality 98A.2; 98B.8, 11; 98E.1; 98G.4–5; 105; 106H; 106I
ḏāt al-ǧiha	modal 98B.10; 105.2
WṢF	
waṣafa bi-	to predicate, to qualify as 98A.12–13,18–19; 105.27
WḌʿ	
mawḍūʿ	subject 98B.13; 98C.7; 98D.2; 98G.8 / posited 100C.6; 105.9,14
mawḍiʿ	topic 126.1–2;132
WFQ	
bi-l-ittifāq	by chance 98B.19

Titles of Ancient Works Referred to in the Texts
(for the titles of works by Theophrastus see **68**)

Persons and Groups Referred to in the Texts

INDICES TO THE COMMENTARY

Texts Discussed or Cited

Persons and Groups (Ancient and Medieval)

(For persons marked * see also Index of Texts Discussed or Cited)

General Subject Index

PHILOSOPHIA ANTIQUA

A SERIES OF STUDIES ON ANCIENT PHILOSOPHY

EDITED BY

K.A. ALGRA, F.A.J. DE HAAS
J. MANSFELD, D.T. RUNIA

13. Nicolaus Damascenus. *On the Philosophy of Aristotle*. Fragments of the First Five Books, Translated from the Syriac with an Introduction and Commentary by H. J. Drossaart Lulofs. Reprint of the 1st (1965) ed. 1969. ISBN 90 04 01725 9
14. Edelstein, L. *Plato's Seventh Letter*. 1966. ISBN 90 04 01726 7
17. Gould, J. B. *The Philosophy of Chrysippus*. Reprint 1971. ISBN 90 04 01729 1
18. Boeft, J. den. *Calcidius on Fate*. His Doctrine and Sources. 1970. ISBN 90 04 01730 5
20. Bertier, J. *Mnésithée et Dieuchès*. 1972. ISBN 90 04 03468 4
21. Timaios Lokros. *Über die Natur des Kosmos und der Seele*. Kommentiert von M. Baltes. 1972. ISBN 90 04 03344 0
23. Iamblichus Chalcidensis. *In Platonis dialogos commentariorum fragmenta*. Edited with Translation and Commentary by J. M. Dillon. 1973. ISBN 90 04 03578 8
24. Timaeus Locrus. *De natura mundi et animae*. Überlieferung, Testimonia, Text und Übersetzung von W. Marg. Editio maior. 1972. ISBN 90 04 03505 2
26. Gersh, S. E. *Κίνησις ἀκίνητος*. A Study of Spiritual Motion in the Philosophy of Proclus. 1973. ISBN 90 04 03784 5
27. O'Meara, D. *Structures hiérarchiques dans la pensée de Plotin*. Étude historique et interprétative. 1975. ISBN 90 04 04372 1
28. Todd, R. B. *Alexander of Aphrodisias on the Stoic Physics*. A Study of the *De Mixtione* with Preliminary Essays, Text, Translation and Commentary. 1976. ISBN 90 04 04402 7
29. Scheffel, W. *Aspekte der platonischen Kosmologie*. Untersuchungen zum Dialog 'Timaios'. 1976. ISBN 90 04 04509 0
31. Edlow, R.B. *Galen on Language and Ambiguity*. An English Translation of Galen's *De Captionibus* (On Fallacies), With Introduction, Text and Commentary. 1977. ISBN 90 04 04869 3
34. Epiktet. *Vom Kynismus*. Herausgegeben und übersetzt mit einem Kommentar von M. Billerbeck. 1978. ISBN 90 04 05770 6
35. Baltes, M. *Die Weltentstehung des platonischen Timaios nach den antiken Interpreten*. Teil 2. Proklos. 1979. ISBN 90 04 05799 4
39. Tarán, L. *Speusippus of Athens*. A Critical Study with a Collection of the Related Texts and Commentary. 1982. ISBN 90 04 06505 9
41. O'Brien, D. *Theories of Weight in the Ancient World*. Four Essays on Democritus, Plato and Aristotle. A Study in the Development of Ideas 2. Plato: Weight and Sensation. The Two Theories of the 'Timaeus'. 1984. ISBN 90 04 06934 8
45. Aujoulat, N. *Le Néo-Platonisme Alexandrin: Hiéroclès d'Alexandrie*. Filiations intellectuelles et spirituelles d'un néo-platonicien du Ve siècle. 1986. ISBN 90 04 07510 0
46. Kal, V. *On Intuition and Discursive Reason in Aristotle*. 1988. ISBN 90 04 08308 1
48. Evangeliou, Ch. *Aristotle's Categories and Porphyry*. 1988. ISBN 90 04 08538 6
49. Bussanich, J. *The One and Its Relation to Intellect in Plotinus*. A Commentary on Selected Texts. 1988. ISBN 90 04 08996 9
50. Simplicius. *Commentaire sur les Catégories*. Traduction commentée sous la direction

de I. Hadot. I: Introduction, première partie (p. 1-9, 3 Kalbfleisch). Traduction de Ph. Hoffmann (avec la collaboration d'I. et P. Hadot). Commentaire et notes à la traduction par I. Hadot avec des appendices de P. Hadot et J.-P. Mahé. 1990. ISBN 90 04 09015 0

51. Simplicius. *Commentaire sur les Catégories.* Traduction commentée sous la direction de I. Hadot. III: Préambule aux Catégories. Commentaire au premier chapitre des Catégories (p. 21-40, 13 Kalbfleisch). Traduction de Ph. Hoffmann (avec la collaboration d'I. Hadot, P. Hadot et C. Luna). Commentaire et notes à la traduction par C. Luna. 1990. ISBN 90 04 09016 9
52. Magee, J. *Boethius on Signification and Mind.* 1989. ISBN 90 04 09096 7
54. Fortenbaugh, W. W., et al. (eds.) *Theophrastes of Eresos.* Sources for His Life, Writings, Thought and Influence. 1992. ISBN 90 04 09440 7 *set*
55. Shankman, A. *Aristotle's* De insomniis. A Commentary. ISBN 90 04 09476 8
56. Mansfeld, J. *Heresiography in Context.* Hippolytos' *Elenchos* as a Source for Greek Philosophy. 1992. ISBN 90 04 09616 7
57. O'Brien, D. *Théodicée plotinienne, théodicée gnostique.* 1993. ISBN 90 04 09618 3
58. Baxter, T. M. S. *The Cratylus.* Plato's Critique of Naming. 1992. ISBN 90 04 09597 7
59. Dorandi, T. (Hrsg.) *Theodor Gomperz. Eine Auswahl herkulanischer kleiner Schriften (1864-1909).* 1993. ISBN 90 04 09819 4
60. Filodemo. *Storia dei filosofi. La stoà da Zenone a Panezio* (PHerc. 1018). Edizione, traduzione e commento a cura di T. Dorandi. 1994. ISBN 90 04 09963 8
61. Mansfeld, J. *Prolegomena.* Questions to be Settled Before the Study of an Author, or a Text. 1994. ISBN 90 04 10084 9
62. Flannery, S.J., K.L. *Ways into the Logic of Alexander of Aphrodisias.* 1995. ISBN 90 04 09998 0
63. Lakmann, M.-L. *Der Platoniker Tauros in der Darstellung des Aulus Gellius.* 1995. ISBN 90 04 10096 2
64. Sharples, R.W. *Theophrastus of Eresus.* Sources for his Life, Writings, Thought and Influence. Commentary Volume 5. Sources on Biology (Human Physiology, Living Creatures, Botany: Texts 328-435). 1995. ISBN 90 04 10174 8
65. Algra, K. *Concepts of Space in Greek Thought.* 1995. ISBN 90 04 10172 1
66. Simplicius. *Commentaire sur le manuel d'Épictète.* Introduction et édition critique de texte grec par Ilsetraut Hadot. 1995. ISBN 90 04 09772 4
67. Cleary, J.J. *Aristotle and Mathematics.* Aporetic Method in Cosmology and Metaphysics. 1995. ISBN 90 04 10159 4
68. Tieleman, T. *Galen and Chrysippus on the Soul.* Argument and Refutation in the *De Placitis* Books II-III. 1996. ISBN 90 04 10520 4
69. Haas, F.A.J. de. *John Philoponus' New Definition of Prime Matter.* Aspects of its Background in Neoplatonism and the Ancient Commentary Tradition. 1997. ISBN 90 04 10446 1
71. Andia, Y. de. *Henosis.* L'Union à Dieu chez Denys l'Aréopagite. 1996. ISBN 90 04 10656 1
72. Algra, K.A., Horst, P.W. van der, and Runia, D.T. (eds.) *Polyhistor.* Studies in the History and Historiography of Ancient Philosophy. Presented to Jaap Mansfeld on his Sixtieth Birthday. 1996. ISBN 90 04 10417 8
73. Mansfeld, J. and Runia, D. T. *Aëtiana.* The Method and Intellectual Context of a Doxographer. Volume 1: The Sources. 1997. ISBN 90 04 10580 8
74. Slomkowski, P. *Aristotle's* Topics. 1997. ISBN 90 04 10757 6
75. Barnes, J. *Logic and the Imperial Stoa.* 1997. ISBN 90 04 10828 9
76. Inwood, B. and Mansfeld, J. (eds.) *Assent and Argument.* Studies in Cicero's *Academic Books.* Proceedings of the 7th Symposium Hellenisticum (Utrecht, August 21-25, 1995). 1997. ISBN 90 04 10914 5
77. Magee, J. (ed., tr. & comm.) *Anicii Manlii Severini Boethii* De divisione liber. Critical Edition, Translation, Prolegomena, and Commentary. 1998. ISBN 90 04 10873 4

78. Olympiodorus. *Commentary on Plato's* Gorgias. Translated with Full Notes by R. Jackson, K. Lycos & H. Tarrant. Introduction by H. Tarrant. 1998. ISBN 90 04 10972 2
79. Sharples, R.W. *Theophrastus of Eresus*. Sources for his Life, Writings, Thought and Influence. Commentary Volume 3.1. Sources on Physics (Texts 137-223). With Contributions on the Arabic Material by Dimitri Gutas. 1998. ISBN 90 04 11130 1
80. Mansfeld, J. *Prolegomena Mathematica*. From Apollonius of Perga to Late Neo-platonism. With an Appendix on Pappus and the History of Platonism. 1998. ISBN 90 04 11267 7
81. Huby, P. *Theophrastus of Eresus*. Sources for His Life, Writings, Thought and Influence. Commentary Volume 4. Psychology (Texts 254-327). With Contributions on the Arabic Material by D. Gutas. 1999. ISBN 90 04 11317 7
82. Boter, G. *The* Encheiridion *of Epictetus and Its Three Christian Adaptations*. Transmission and Critical Editions. 1999. ISBN 90 04 11358 4
83. Stone, M.E. and Shirinian, M.E. *Pseudo-Zeno. Anonymous Philosophical Treatise*. Translated with the Collaboration of J. Mansfeld and D.T. Runia. 2000. ISBN 90 04 11524 2
84. Bäck, A.T. *Aristotle's Theory of Predication*. 2000. ISBN 90 04 11719 9
85. Riel, G. Van. *Pleasure and the Good Life*. Plato, Aristotle, and the Neoplatonists. 2000. ISBN 90 04 11797 0
86. Baltussen, H. *Theophrastus against the Presocratics and Plato*. Peripatetic Dialectic in the *De sensibus*. 2000/ ISBN 90 04 11720 2
87. Speca, A. *Hypothetical Syllogistic and Stoic Logic*. 2001. ISBN 90 04 12073 4
88. Luna, C. *Trois Études sur la Tradition des Commentaires Anciens à la Métaphysique d'Aristote*. 2001. ISBN 90 04 120074 2
89. Frede, D. & A. Laks (eds.) *Traditions of Theology*. Studies in Hellenistic Theology, its Background and Aftermath. 2001. ISBN 90 04 12264 8
90. Berg, R.M. van den. *Proclus' Hymns*. Essays, Translations, Commentary. 2001. ISBN 90 04 12236 2
91. Rijk, L.M. de. *Aristotle – Semantics and Ontology*. 2 volumes.
    Volume I. General Introduction. The Works on Logic. 2002. ISBN 90 04 12324 5
    Volume II. The Metaphysics. Semantics in Aristotle's Strategy of Argument. 2002. ISBN 90 04 12467 5
92. Finamore, J.F. & J.M. Dillon. *Iamblichus* De Anima. Text, Translation, and Commentary. 2002 ISBN 90 04 12510 8
93. Fortenbaugh, W.W., R.W. Sharples, & M.G. Sollenberger. *Theophrastus of Eresus*. On Sweat, on Dizziness and on Fatigue. 2003. ISBN 90 04 12890 5
94. Tieleman, T. *Chrysippus'* On affections. Reconstruction and Interpretation. 2003. ISBN 90 04 12998 7
95. Görler, W. *Kleine Schriften zur hellenistisch-römischen Philosophie*. Herausgegeben von C. Catrein. 2004. ISBN 90 04 13736 X
96. Polito, R. *The Sceptical Road*. Aenesidemus' Appropriation of Heraclitus. 2004. ISBN 90 04 13742 4
97. Fortenbaugh, W.W. *Theophrastus of Eresus*. Sources for his Life, Writings, Thought and Influence. Commentary Volume 8. Sources on Rhetoric and Poetics (Texts 666-713). 2005. ISBN 90 04 14247 9
98. Perkams, M. & R.M. Piccione (Hrsg.) *Proklos. Methode, Seelenlehre, Metaphysik*. Akten der Konferenz in Jena am 18.-20. September 2003. 2006. ISBN 90 04 15084 6
99. Schäfer, C. *Philosophy of Dionysius the Areopagite*. An Introduction to the Structure and the Content of the Treatise *On the Divine Names*. 2006. ISBN 90 04 15094 3
100. Delcomminette, S. *Le* Philèbe *de Platon*. Introduction à l'Agathologie Platonicienne. 2006. ISBN 90 04 15026 9
101. Fortenbaugh, W.W. *Aristotle's Practical Side*. On his Psychology, Ethics, Politics and Rhetoric. 2006. ISBN-10: 90 04 15164 8, ISBN-13: 978 90 04 15164 2

102. Brancacci, A. & P.-M. Morel (eds.). *Democritus: Science, The Arts, and the Care of the Soul.* Proceedings of the International Colloquium on Democritus (Paris, 18-20 September 2003). 2007.
ISBN-10: 90 04 15160 5, ISBN-13: 978 90 04 15160 4
103. Huby, P. *Theophrastus of Eresus.* Sources for his Life, Writings, Thought and Influence. Commentary Volume 2. Logic. With Contributions on the Arabic Material by Dimitri Gutas. 2007.
ISBN-10: 90 04 15298 9, ISBN-13: 978 90 04 15298 4
104. Sider, D. & C.W. Brunschön (eds.). *Theophrastus of Eresus.* On Weather Signs. 2007.
ISBN-10: 90 04 15593 7, ISBN-13: 978 90 04 15593 0